A DICTIONARY OF ETHOLOGY

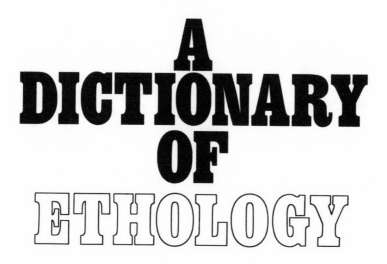

A DICTIONARY OF ETHOLOGY

Klaus Immelmann / Colin Beer

HARVARD UNIVERSITY PRESS
Cambridge, Massachusetts
London, England
1989

Library of Congress Cataloging-in-Publication Data

Immelmann, Klaus.
  A dictionary of ethology.

  Translation of: Wörterbuch der Verhaltens-
forschung.
  Bibliography: p.
  1. Animal behavior—Dictionaries.   I. Beer,
Colin. II. Title.
QL750.3.I4513 1989   591.5′1′0321   88-21360
ISBN 0-674-20506-5 (alk. paper)

To the memory of Klaus Immelmann

CONTENTS

PREFACE

Every branch of biology, indeed every branch of science, has had to create some technical terms for precision in communication and rigor in argument. The relatively young science of ethology—the biological study of animal behavior—is no exception. And like the other sciences, it has had to contend with the difficulty that when an objective description is needed, the terms that seem most appropriate have connotations that may lead to misinterpretation. This problem has been especially serious in ethology, where the temptation to make hasty, sometimes anthropomorphic interpretations is greater than in the other life sciences. Consider how easily the behavior of animals can suggest mother love, courage, stupidity, wickedness, or cleverness—ascriptions that can rarely be applied literally to animals in the senses that apply to humans.

The difficulties that ethology has had in developing a technical language can be understood historically. When the study of animal behavior began, two sources of terminology were already available: everyday language and the jargon of related sciences such as anatomy, ecology, and, especially, psychology. Ethology adopted and adapted words from both sources. However, as work in the field progressed, it became clear that many of the words that apply to both animals and humans either have different meanings in the two contexts or involve unwarranted interpretation concerning animals. Nevertheless, once a term was admitted to the ethological vocabulary, it tended to stay. Consequently, numerous words used in animal behavior study have meanings different from those that apply in common parlance. Prime examples are "information," "display," "dialect," "learning," "social," and the controversial "instinctive." Similar differences obtain between ethology and psychology; "displacement," for example, does not mean the same thing to an ethologist as it does to a Freudian psychoanalyst. Finally, even within the discipline, many technical terms are understood differently by the different schools of animal behaviorists.

This dictionary is designed to serve several purposes. Its main function is to explicate the more important terms used in the study of animal behavior and to survey their range of application. In addition, for readers with little advanced knowledge of biology, it defines biological terms that are frequently used in ethological writing. Finally, it tries to situate the technical terms used in ethology and other subdisciplines of biology within a framework of general biology, showing their special importance for behavior study and the relationship of their behavioral significance to their meanings in other contexts.

The selection of terms posed some difficult problems. On the one hand, we did not want to overload the book with highly specialized terms likely to be of interest only to an expert. On the other hand, we wanted to provide a comprehensive working vocabulary sufficient for understanding ethological modes of inquiry and patterns of argument. Accordingly we excluded highly technical terms that apply only to one or a few species and therefore serve the needs mainly of specialists. The number of these specialized terms is so great that their inclusion would have added bulk far in excess of what is reasonable for a general reference book. The only exceptions to this exclusion are those specialized terms that have some significance for broader interests, either because they played a part in the history of ethology—such as the triumph ceremony of the greylag goose and the egg rolling of ground-nesting birds—or because they serve to illustrate or illuminate some more general ethological principle.

We have included a number of terms from neighboring disciplines to which there are important bridges and from which ethology has borrowed. Of these disciplines, the most drawn on are evolutionary biology and ecology, especially sociobiology, the branch recently developed from their conjunction. Terms from sensory physiology and neurophysiology are included but to only a limited extent because they already have their own detailed dictionaries. Also the more biological approaches to human psychology have drawn progressively closer to ethology in recent times, resulting in a fruitful exchange of concepts.

For some time now opinion has been divided in ethology about which technical terms are really necessary, which are dispensable, and which are perhaps even pernicious because their vagueness or ambiguity stands in the way of precise definition. Controversy has gathered around such terms as "motivation," "imprinting," "play behavior," and "rank order." This book includes a number of these disputed terms, not to defend the use of particular concepts, such as imprinting, but to promote better communication among specialists who are vaguely aware of the problematic nature of such concepts and who may welcome a ready source of reference when they encounter the concepts in use. For terms that are or have been controversial or are used in more than one sense, we have taken pains to explain the nature of the problem, to distinguish the differences of meaning, and to chart the range of application.

In spite of the careful consideration given to the selection of terms our compilation is no doubt far from definitive. Readers may note the preponderance of higher vertebrates in the examples chosen to support the definitions. This is more than a reflection of our own research interests, which have been mainly in the area of bird behavior; it is a consequence of the fact that, from the beginning, observations and experiments on birds and mammals have supplied a high proportion of ethology's basic ideas and technical concepts. Thus it is possible that many of our explanations of concepts will prove to be incomplete or open to challenge. We cannot claim that our formulations are free of personal views and preferences, Indeed, we were not in agreement on all issues.

The present dictionary has three predecessors, written in German by Klaus Immelmann: a selection of definitions of ethological terms compiled for the supplementary volume *Verhaltensforschung* of the encyclopedic series *Grzimek's Tierleben,* published by Kindler-Verlag in 1974; an expanded compilation, *Worterbuch der Verhaltensforschung,* a volume in the series *Geist und Psyche,* published in 1975 also by Kindler-Verlag; and a still further enlarged edition with the same title, published by Verlag Paul Parey in 1982. The present book is in part a translation of the 1982 German edition. However, some of the German terms have no precise equivalents in English, and some English terms are not represented in the German list. Even when there are corresponding terms in the two languages, their connotations sometimes differ in detail. Consequently the selection of terms in this English version departs appreciably from that in the German version, and some of the translation is very free.

This dictionary has the following format. Each entry begins with any words that are synonymous or closely related in meaning to the term being defined. These provide an immediate capsule definition, especially for an unfamiliar or esoteric term for which another English word is equivalent or used in a similar way. Next comes a brief explication of the term, followed by examples, either particular instances or an enumeration of the animal groups in which the behavior occurs. In the case of the many terms that are used differently by different authors or by different schools, the various uses are set forth. Occasionally someone else's definition is quoted verbatim (or translated as literally as possible), in which case the name of the author is given.

A large number of terms are listed with only a cross-reference to another term. This does not necessarily mean that the two terms are synonymous. On the contrary, only a broad similarity of sense may justify the joint explication of a whole group of concepts, or two terms may even be exact opposites, in which case they are dealt with together for the sake of economy of space.

Many people contributed to the realization of this book. The late Klaus Immelmann had wanted to acknowledge the stimulation, advice, and criticism of his colleagues at the University of Bielefeld, and especially to thank H. J. Bischoff, P. Hammerstein, H. Hendrichs, E. Pröve, N. Sachser, P. Sonnemann, R. Sossinka, K. Weigel and R. Witt for their help with the German text. In addition, I am in debt to Ernst Mayr and W. John Smith, who read the manuscript and provided useful comments; to Thomas Immelmann, who took care of the necessary arrangements after his father's death; to Laszlo Meszoly, who drew the illustrations; and to those people at Harvard University Press, especially Howard Boyer, who promoted the completion of the book and saw it through the acceptance process, and to Peg Anderson, whose editorial skill and stylistic taste clarified and smoothed the text in many places.

As the manuscript of this book was nearing completion, Professor Immelmann died suddenly at his home in Bielefeld, West Germany. May the book help to preserve our memories of him and to sustain his profound influence on the science of ethology.

A DICTIONARY OF ETHOLOGY

AGGRESSION

Abnormal behavior Behavioral anomaly. Any behavior deviating from the norm. In practice the concept has only marginal value, since it is usually very hard to say what behavioral characteristics constitute the normal behavioral repertoire of a species and which are to be considered deviant. For example, should the stereotyped routines of caged animals continue to be classed as abnormalities when such behavior is found in their free-living conspecifics?

Besides the behavioral consequences of disease, illness, and aging, abnormal behavior includes fixation on inappropriate objects (see ERRONEOUS IMPRINTING), deficits due to lack of specific social stimulation (see DEPRIVATION SYNDROME), and compulsive movement patterns consequent on unnatural living conditions (see STEREOTYPY).

Acoustic isolation See Isolation experiment

Acoustic window See Melotope

Action See Behavior

1

Action chain   Reaction chain. A stereotyped sequence by two animals in which a response by one presents the stimulus for the other's next act, which, in turn, stimulates the first animal to further action, and so on. Although the order of the sequence usually does not vary, the degree of fixity or flexibility varies considerably from case to case. Action chains are seen in such social contexts as courtship (courtship chains) and ritualized fighting. A frequently cited example of a courtship chain, and one of the first to be investigated in the history of ethology, is the pairing behavior of the three-spined stickleback, in which male and female alternate the roles of actor and reactor. Since such a succession of interactions decreases the likelihood of mistakes in mating partnerships, courtship chains can contribute to reproductive isolation between different species.

Such socially mediated action chains bear some analogy to chain reflexes, in which activation of one reflex causes stimulation that activates another, which triggers a third, and so on, within the same organism. Attempts have been made to account for the patterning of locomotory movement in animals in this way. Operant conditioning can also produce the chaining of responses in a single animal, as is frequently done in the training or shaping of the performances of circus animals.

Action potential   In biology this term is used in two connections: in neurophysiology it is synonymous with nerve impulse; in now more or less obsolete ethologese it meant the probability of occurrence of a specific behavior pattern. The ethological usage grew out of attempts to deal with behavioral variation that is not attributable to peripheral stimulation (see THRESHOLD CHANGE; VACUUM ACTIVITY). Some behavior patterns are easier to elicit with increasing time after their last occurrence and in the absence of any corresponding change in sensory input, which seems to imply a build-up of internal factors responsible for the patterns. Lorenz (1950) postulated spontaneous generation of "action-specific energy" (ASE) for each such behavior pattern. Other ethologists judged that the energy model assumed more than was warranted by or consistent with the evidence, so the less loaded term "specific-action potential(ity)" (SAP) came to be preferred to ASE. Generally SAP meant the readiness for elicitation of a behavior pattern without any substantive implications about its underlying physiology, but for some ethologists the term

2

continued to have the connotation of causal agency. In its purely descriptive or predictive sense, SAP has been replaced by "behavioral" or "response tendency" or "probability" in the writing of most ethologists. However, some, such as Lorenz, continue to use "action potential" in much the same sense as the earlier concept of motivational energy and continue to defend the essentials of the hydraulic model of motivation of which it was a part.

Action-specific energy   See Action potential; Motivational energy

Action-specific exhaustibility   Behavioral fatigue. Refractoriness of certain behavior patterns to repeated stimulation: after the behavior has been performed once or a few times, it cannot be elicited again for some time, or only by much stronger stimulation than was sufficient initially. The fatigue of one pattern generally leaves others unaffected; the refractoriness is specific to the action. Behavior patterns differ in the degree to which they show the effect. Thus fleeing and defensive reactions, as a rule, can be elicited again immediately after performance, in contrast to sexual and feeding activity, which may require a lengthy pause before repetition. A classical example of an almost inexhaustible activity is the retrieval of pups by rodents. The action specificity of the exhaustion shows that muscle fatigue can be ruled out as its cause, for the same muscles may be involved in unaffected patterns. In some cases where response to repeated stimulation wanes, the effect has been found to be specific to the stimulus rather than to the action; even a slight variation of the stimulus sometimes is sufficient to reinstate the behavior fully (see COOLIDGE EFFECT).

Activity   Movement in general, or some kind of action in particular (for example, nest-building activity). "An animal is active when it moves parts of its body or moves itself" (Aschoff 1954). Besides the functionally specialized categories of activity, there are more open-ended forms such as curiosity behavior and play behavior.

In most species, activity alternates with rest in some cyclic rhythm, which may consequently be described as a diurnal, nocturnal, or crepuscular rhythm. While many species have a twenty-four-hour activity cycle, others have two or more activity periods per day, and still others have activity cycles pegged to the tidal rhythm. (See CIRCADIAN RHYTHM.)

Measures of general activity are employed in a great deal of behavioral investigation (see OPEN-FIELD TEST).

Adaptation Adjustment to circumstances, accommodation. The word can refer to either the process or its product. Thus we speak of gradual adaptation to changes of climate, and we describe coat qualities as adaptations to such changes. Several other differences of usage are found in ethology.

As in ecology and evolutionary biology, in ethology adaptation is usually taken to be the acquisition of characteristics that make an organism better suited to its current environment or way of life and so maximize the life expectancy and reproductive success of the organism and its descendants. Adaptations thus arise in morphology, physiology, and behavior. Behavior patterns are adapted to an animal's environmental conditions in the same way its other characteristics are. The study of behavioral adaptation is a main concern of ecoethology (see SOCIOBIOLOGY).

Each adaptation requires adequate information about the relevant environmental conditions. The organism can acquire this information either through heredity or through individual experience. In the first instance one speaks of phylogenetic or evolutionary adaptation, since the adaptation must have occurred in preceding generations and been passed on genetically. It occurs in the familiar way through mutation and selection; the individuals that leave the most descendants are those whose genes led to development of the most favorable characteristics in their environmental conditions. In this way such characteristics permeate more and more of the population as generation follows generation. Evolutionary adaptation thus consists in change of frequency of characteristics in a population.

Individual experience is the other means of effecting adaptation, which can be called individual adaptation. By this means individuals or populations can develop different behavioral characteristics, that depend upon the environment in which they grow up and to which they must adjust. As a rule both adaptive processes work together, evolutionary adaptation determining the frame within which individual adaptation can be exercised. Thus evolution has fashioned a baboon into a creature that seeks a place elevated above the ground

4

when settling down for the night; whether it chooses a cliffside or a tree will depend upon its individual experience. In higher vertebrates learned behavioral adaptation can sometimes be passed on from one generation to the next (see TRADITION).

Besides the ecological and evolutionary senses, adaptation covers other biological phenomena. In sensory physiology, for example, sensory adaptation refers to the adjustment of a sense organ or receptor cell to the prevailing level of stimulation, as can be experienced visually when going from bright sunlight into a darkened room or vice versa. Another form of physiological change called adaptation is the reduction of excitability of a sense organ or sense cell after continuous or repeated stimulation. Such change in sensitivity requires that a persistent or repetitive stimulus be progressively increased in strength to produce the level of response initially elicited. This is a temporary adjustment process, to be distinguished from sensory fatigue on the one hand (which is not stimulus specific, as the adaptation process is) and habituation on the other. However, habituation—learning not to respond to a stimulus that nothing is contingent on—is also sometimes referred to as adaptation. Since, in this case, the change results from central rather than peripheral adjustments, it is sometimes described as central nervous system adaptation.

Adaptive correlation   Association between form and function; preciseness of fit between the characteristics of species and the natural selection pressures to which their habitats and ways of life expose them. Also, the study that reveals association and preciseness of fit. In ethology the best-known example of adaptive correlation is the comparison of cliff-nesting kittiwake gulls and the more typical ground-nesting gulls. Kittiwakes do not have to contend with the predator pressures to which ground-nesting gulls are exposed, but they are beset by space limitations on the cliff ledges and the hazard to eggs and young of falling out of the nest. The breeding behavior of the two kinds of gulls differs in a number of ways correlated with these differences in the problems presented by the breeding habitat.

Adaptive radiation   Splitting up and divergence of the descendants of a taxonomic group. It is a phylogenetic development consisting of multiple branching of a lineage into species and genera adapted to very different environmental conditions or ways of life. Such a divergent splitting up of forms is especially likely where one, or a few, pioneer forms colonizes a new habitat having a large number of empty ecological niches and no competitors. Examples of rapid adaptive radiation are the Australian marsupials, the cichlid fishes of East African lakes, Darwin's finches on the Galapagos Islands, and the Hawaiian honeycreepers. In a geologically rather short time these groups have brought forth remarkably various forms. The closely related species of Darwin's finches include seed eaters, insect eaters, nectar eaters, and fruit eaters—feeding specializations that, as a rule, are characteristic of higher taxonomic units (families, sometimes orders). Behavior can be of special significance in adaptive radiation, for a change in behavior frequently constitutes a first step that can lead to changes in other characteristics, such as body form, in an evolutionary adaptive shift.

Adaptive significance   See Survival value

Adequate stimulus   See Receptor; Stimulus specificity

Adjunctive behavior   See Displacement activity

Adoption   In ethological writing, any situation in which individuals take over the care of young that do not belong to them and act more or less like full-time substitute parents; merely temporary or casual aid, such as that contributed by helpers and allomothers, does not qualify. The concept thus includes the taking over of young by close relatives, even immediate kin; it appears that in animal society adoption is more frequently contracted between close relatives than between distantly related or unrelated individuals, which is in agreement with the predictions of kin selection theory. Thus in chimpanzees and rhesus monkeys the adopters are often older sisters, who, after the death of the mother, are intensely concerned with the orphaned infant. Adoption of young by relatives has also been observed in other primates (macaques, langurs), in wild dogs, and in coatis.

In contrast, animals living in groups in which there is probably no kin recognition allow orphaned young to perish in most cases. An exception to this may occur in the creche formations of many

6

birds (see CRECHE), where the adults in attendance at any one time are unrelated to most of the young. This is possibly an instance of reciprocal altruism.

A special form of adoption has been observed in hamadryas baboons, in which young adult males abduct sexually immature females from the harem of an older male, thereby establishing the beginning of what will later be their own harem.

Adult Fully grown, sexually mature (where appropriate); an animal so described.

Advertisement See Display

Advertising dress In cases of sexual dimorphism, one sex is often more conspicuous in form and coloration than the other. Usually it is the male that is dressed for advertisement. Exceptions to this are the phalaropes and seed snipe. The conspicuous appearance may be permanent or may alternate with an inconspicuous phase in synchrony with the seasons. When the advertising dress is paraded only during the breeding period, it may be referred to as nuptial dress or nuptial coloration. Examples are the dorsal fin serrations of many male salamanders, the red coloration of three-spined stickleback males, and the brightly colored plumage of the drakes in many species of ducks. Male birds that molt to an inconspicuous appearance outside the breeding season are said to be in eclipse plumage. Features of advertising dress commonly serve as releasers in reproductive and agonistic contexts.

Afference Nervous excitation conveyed from sense organs (including proprioceptors and visceral receptors) to the central nervous system. The nerves carrying this excitation are called afferent nerves or sensory nerves, in contrast to those carrying excitation from the central nervous system to muscles and glands, which are called efferent nerves or motor nerves (see EFFERENCE).

Ethology draws an important distinction between stimulation brought about through the animal's own movement (see REAFFERENCE) and stimulation arising from external causes with respect to which the animal is passive (see EXAFFERENCE). Thus if an animal turns its head, the consequent change in retinal input is reafference;

if the animal stays still and another moves across its line of sight, the resulting visual stimulation is exafference. This distinction has been of value in elucidating mechanisms of sensory-motor control in animals.

Age dimorphism   Difference of form between young and adult animals. In many animal species individuals below a certain age or stage of development differ in appearance from older or more mature individuals. Juvenile characteristics serve various behavioral functions, including camouflage for defense against predators that present no danger to adults and signals to induce adults to act benignly toward them. Age dimorphism is especially pronounced in species that go through a larval stage of development. When more than two developmental stages are distinguished by appearance, the case is described as one of age polymorphism.

Aggregation   Congregation of animals that does not depend upon social attraction, that comes about because a number individuals are singly seeking the same place. Aggregations form at watering places, in good hiding places, and in places suitable for sleep or hibernation. For certain species of birds there are clear indications that resort to a secure night roost where many conspecifics gather can have a further, socially provided benefit. The roost can serve as an information center; individuals that have found a good supply of food will depart early and conspicuously the next morning, stimulating the previously less successful foragers to follow them to the plentiful place. This case verges on true social grouping, in which animals congregate because they are attracted to one another. Since it is on the boundary of "true" social grouping, aggregation is sometimes described as subsocial grouping.

Aggression   Aggressive behavior. A general term for all elements of attack, defense, and threat behavior. Among several distinctions that have to be drawn, perhaps the clearest is between intraspecific and interspecific aggression—altercations involving members of the same species and those involving members of different species. In the interspecific category one has to distinguish between the prey-catching assault of predators on their intended victims, defensive resistance of victims against predators, and struggles between competitors of different species over possession of or access to a contended resource. As a rule interspecific aggression refers to competitive rather than predator-prey interaction. Intraspecific and interspecific aggression differ in the type of fighting usually employed: between members of the same species, ritualized fighting is typical; in interspecific aggression injurious fighting is more common. The biological significance of intraspecific aggression has frequently been discussed in the ethological literature. There is little doubt that it contributes to the establishment of adaptive social organization according to species type; that is, the animals of a population, singly, in pairs, or in groups, divide up their habitat more or less evenly so that they utilize it to the best of their ability. If population density exceeds the load the habitat can carry, aggression may force the surplus individuals to go elsewhere before food shortage extensively weakens the population. Aggression can also be useful in sexual selection and in the formation of social hierarchy. From the point of view of population biology, aggression is seen as "behavior actively directed against competitors, which decreases the reproductive prospects of the attacked animal relative to those of the attacker, for example by denying access to certain resources" (Markl 1976).

In addition to these types of aggression, experimental studies of captive animals have contributed frustration-induced and pain-induced aggression. Still other forms arise in human society, such as military and economic aggression between nations.

Aggressiveness A species' or individual's typical tendency to attack, or a tendency specific to an individual on a particular occasion or in a particular situation. For each species aggressiveness lies within certain limits, which can fluctuate with the breeding condition, and it may differ widely between closely related species. Within these limits environmental factors and experience can affect an individual's level of aggressiveness. Still unsettled questions are whether there is, at least in some species, an autonomous aggressive drive, that is, whether (and if so, in what way) a set of internal factors determines an individual's readiness to fight and to what extent aggressiveness can be "dammed up" or erupt spontaneously (*in vacuo*). (See DRIVE; MOTIVATION; SPONTANEOUS BEHAVIOR; VACUUM ACTIVITY.)

9

Agonistic behavior A general term for all behavior having to do with fighting, including aggression (attack, threat behavior, defense) and fleeing. The use of the term in the ethological literature varies: many ethologists, especially those writing in English, treat agonistic behavior as synonymous with aggressive behavior, thus excluding fleeing. But, in most animal fights, elements of attacking and fleeing are closely combined, so there is a need for a general term covering both. A clear distinction between agonistic and aggressive behavior is therefore preferred.

Another distinction to be noted is that in neuroendocrinology and psychopharmacology the word "agonist" refers to a drug that can substitute for a hormone or neurochemical substance at neural receptor binding sites, producing the same effects as the natural substance or enhancing its effectiveness. This is in contrast to an antagonist, a drug that likewise binds to the receptor site but in so doing blocks the effect of the natural substance.

Agonistic buffering A concept recently formulated in the primate literature to refer to interaction between two males in which one carries an infant. The infant is used as an aggression-inhibiting tool in encounters that would otherwise involve a high probability of a hostile clash. In this way the two males may spend several hours in peaceful social contact. Agonistic buffering has been observed mostly in the multi-male groups of macaques and baboons. A similar use of children to signal friendly intention is known from numerous human cultures, for example the Waika Indians and the Australian aborigines.

Alarm call See Warning behavior

Alarm substance See Warning behavior

Allele See Gene

Allochthonous drive In connecton with displacement activity, the surplus hypothesis postulates that irrelevant behavior is caused by motivational energy displaced from the activities thwarted or in conflict. For example, if an attack drive is obstructed by fear it may be expressed as grooming. The displacement behavior is thus an expression of an allochthonous drive—a source of motivation other than its own. When caused by its own drive, the behavior is said to be autochthonously driven. The terms were introduced by Kortlandt (1940).

Allogrooming See Grooming; Social grooming

Allomarking See Scent marking

Allomother Foster mother. A term so far applied only to female primates that show parental behavior (retrieving, guarding, cleaning, and so on—see PARENTAL CARE) toward infants not their own. In earlier primate literature such temporary stepmothers were referred to as aunts, but this has been objected to because the implied degree of relationship does not always obtain. As in the case of helping in other animal groups, the biological significance of this seemingly altruistic behavior is not entirely clear (see ALTRUISM). One direct benefit could lie in the accumulation of experience useful to the allomother when she later has young of her own to care for. If she is closely related to the natural mother, the gain in inclusive fitness presents another advantage (see KIN SELECTION THEORY). The infants receiving the attention may come to know members of the group sooner than they otherwise would, thus putting at their disposal a greater number of familiar social companions to whom they can flee when frightened. This could also be of indirect advantage to the real mother. As yet there are so few pertinent field observations that these remarks are only conjectures. In some primate species males temporarily take charge of an infant, but in many cases the objective appears to be to decrease agonistic tension (see AGONISTIC BUFFERING).

Allopatric Geographically isolated. Because there can be no gene flow between populations of the same species separated by geographical barriers that prevent interchange of individuals, such populations may diverge genetically to the point where interbreeding either would be impossible or would lead to poor reproductive success should the barriers be removed. Thus geographic isolation can lead to allopatric speciation (compare SYMPATRIC).

Allopreening   See Social grooming

All-or-nothing responsiveness   The situation in which stimulation either is sufficient to evoke a response in its complete form or evokes no response. Thus a complete nerve impulse occurs when the nerve cell membrane is acted on in such a way as to reduce the membrane potential to the threshold level. In contrast, the excitatory postsynaptic potential leading to triggering of the impulse is a graded response, correlated with the amount and kind of synaptic transmission. Simple reflexes typically are of the all-or-nothing type.

Altricial animals   Species in which the young emerge at a relatively early developmental stage (compared with precocial animals). Such young typically lack fur or feathers and are blind and incapable of supporting their own weight and hence incapable of locomotion. They require considerable care from their parents over an extended period, especially in the provision of food and warmth. In birds this period of dependence is spent in the nest, so such nestlings are also described as nidicolous. Examples of altricial birds are pigeons and songbirds; of mammals, most rodents and carnivores.

Altruism   Unselfish behavior. Behavior in which an animal, without regard to its own welfare, benefits another member of its species. Altruism in this broad sense attaches to all parental behavior. It is rarer between adult animals, although the number of confirmed instances has risen considerably since sociobiologists stirred up interest in the question. Its occurrence is associated particularly with three kinds of activity: reciprocal warning and defense, giving of aid in the rearing of young, and sharing of food.

Explaining altruistic behavior has long been difficulty, since such behavior appears to be inconsistent with natural selection, which should favor only selfish behavior. Recently this gap in understanding has been closed, largely through the contributions of sociobiology. It has been shown that much apparently altruistic behavior can also bring benefit to the animal performing the action (see WARNING BEHAVIOR). In other cases, especially when the altruism is unilateral, the giver of help is closely related to the receiver and so gets indirect return through increase in inclusive fitness (see KIN SELECTION THEORY). Altruism between genetically unrelated animals is, as a

12

rule, mutual and implies a more or less equal benefit for the two animals. This relationship is sometimes called "mutualistic," which is not recommended because the term already refers to certain living arrangements, such as symbiosis, involving individuals of different species. Altruism biased toward close kin is sometimes described as nepotism.

Ambivalent behavior Behavior patterns composed of parts of two different, often antagonistic, actions occurring simultaneously or in rapid alternation. Ambivalent behavior can occur when two behavioral tendencies are activated simultaneously (see MOTIVATION). The components are often incompletely performed and may have the character of intention movements. Ambivalent behavior is characteristic of conflict situations and is thus a common ingredient of threat behavior, in which elements of attacking and of fleeing may overlap or alternate with one another. Ritualization of ambivalent behavior is believed to be one of the main ways threat displays have evolved. Examples include the upright posture of gulls (simultaneous action) and the zig zag dance of male three-spined sticklebacks (alternating action).

Ameslan A portmanteau word from American Sign Language. This visual sign system has been used to establish something approaching linguistic communication between people and specially schooled chimpanzees (see LANGUAGE).

Amplitude modulation Variation in the intensity of a signal, for example, fluctuation in the loudness of a sound.

Anachoresis See Defensive behavior

Analogy In phylogenetic comparison, similarity due to parallel or convergent evolutionary adaptation (see CONVERGENCE). Resemblance in body form or behavior, for example, as a consequence not of common ancestry (compare HOMOLOGY), but of independent adaptation to the same or similar environmental conditions. Examples are the dorsal fins of sharks and dolphins, the lens eyes of vertebrates and cephalopods, and, in the context of behavior, the echolocation of bats and oilbirds and the distraction displays of plover and some ducks. Analogous features are said to be analogues of one another.

When ethologists speak of analogy in discussing or presenting theory, it can overlap in meaning with model as used in similar contexts. Analogy obtains in this sense when one subject is likened to another or when two subjects share some features of form or functioning and it is argued that the similarity may extend to other features. For example, the motivational theories of classical ethology were based in part on analogies between behavioral regulation and hydraulic systems; cognitive scientists are now exploring the extent of the analogy between how minds work and how computers work. Although argument by analogy has frequently added useful ideas to science, it has equally frequently led scientists astray and so is often decried as a sort of fallacy.

Androgen   Generic term for male sex hormone, the main source of which in vertebrates is the interstitial cells of the testes. The most common androgen is testosterone.

Anestrus   In mammals, the phase of the reproductive cycle following estrus, when the female is sexually unreceptive and incapable of being fertilized.

Animal communication   See Communication

Animal language   See Language

Animal psychology   A term that has to a large extent been replaced by "ethology," which covers much of what used to be called animal psychology. Ethology has come to be preferred perhaps because of its biological connotation. However, some kinds of animal behavior study that stem historically from human psychology, especially the old school of associationism are still more appropriately described as animal psychology, for example parts of comparative psychology and animal cognition (compare COGNITIVE ETHOLOGY). Today the term is more likely to be used in connection with the personality characteristics of animals—their individuality, subjectivity (insofar as this can be ascertained), and pathological or abnormal behavior—than in connection with their normal behavior. Much of the understanding of animal psychology in this narrower sense comes from observation of zoo and circus animals (see ZOO BIOLOGY). However, most popular writing about animal behavior uses the term "animal psychology" in a very general and loose way.

Animal society   See Society

Animal sociology   A part of ethology concerned with social structure. To the extent that it can be distinguished from sociobiology, it concentrates on the mechanisms that establish and maintain social structures, such as the form of communication, rather than on the selective factors and adaptive significance accounting for the evolution of those structures.

Annual periodicity   Annual rhythm, circannual rhythm. Changes in behavior and in body functions that recur at twelve-month intervals. Most notable are the yearly fluctuations in reproduction observed in most animals, for example, the onset and waning of gonadal activity and the associated appearance and disappearance of secondary sex characteristics. Other annually repeating events are the migrations of many birds, winter hibernation, and the molting cycles and associated color changes of many Arctic mammals and birds. Behavior includes many examples of annual rhythms, especially activities having to do with reproduction, such as courtship, territoriality, nest-building, and parental behavior. Aggressive behavior may also fluctuate in the course of a year; as a rule it peaks at the beginning of the reproductive period.

Investigations have shown that annual periodicity, like daily rhythms, may be governed by an endogenous periodicity. However, this is not so generally the case, or so obvious, as in day-night rhythmicity (see CIRCADIAN RHYTHM). Indeed many annually periodic processes appear to be driven more or less directly by environmental forces such as day length. By analogy with the term "circadian" for daily periodicity, the term "circannual" was coined for annual periodicity.

Anonymous group   An animal group in which the members do not recognize one another individually, in contrast to an individualized group, which does have individual recognition. There are two kinds

15

of anonymous groups: in open groups the members are interchangeable, and the group composition may therefore be in continual flux; in closed groups the members are distinguished from nonmembers by certain group-specific characteristics such as group scent or hive odor. Open groups include insect swarms, fish schools, bird flocks, and the nomadic herds of various mammals. Closed groups include swarms of bees, colonies of ants and termites, and colonies of many rodents. However, even the open anonymous group differs from aggregation in being a product of social attraction.

Antagonism Opposition between effects, processes, or tendencies. In physiology the term applies to the situation in which one chemical, the antagonist (for example, an antiandrogen), interferes with the action of another; and also to muscles that have opposite effects at the same joint (for example, flexors and extensors), which are consequently called antagonistic muscles. In behavior there can be antagonism between opposing tendencies, for example attacking and fleeing (see CONFLICT).

Anthropology The study of mankind as a natural and social entity. It comprises physical anthropology, which deals with such matters as racial physique, technology, and accommodation to and exploitation of the natural environment; and cultural anthropology, which deals with human institutions, including language, religion, and art. Hence anthropology overlaps with many other branches of human study, such as archaeology, history, linguistics, economics, sociology, and political science, as well as with parts of biology, such as anatomy and physiology, ecology, and ethology. The ethological connections are particularly obvious in primate ethology and human ethology.

Anthropomorphism The tendency to attribute human qualities feelings, and capacities to animals or inanimate objects. Animal behavior offers much opportunity for anthropomorphic interpretation implying consciousness, intentionality, sometimes even understanding, especially when observed in higher vertebrates, and above all in cases of a close personal association between an animal and a human keeper. As a rule, such attributions cannot be tested by objective investigation, so it is usually impossible to decide whether they are truly applicable and within what limits. A working principle

16

to counter excessive or cavalier indulgence in anthropomorphism is Lloyd Morgan's Canon. This matter poses hard philosophical as well as practical questions, which have recently received a revival of interest after many years of behavioristic banishment (see BEHAVIORISM; COGNITIVE ETHOLOGY). In the past, anthropomorphic assumptions frequently hindered the discovery of natural ways to maintain animals under human care.

Antihormone   A synthetic substance that blocks the functioning of a hormone. Antihormones have been used most extensively in research on sex hormones, for which they are steroids in most cases, like the sex hormones themselves. They can be used for chemical castration, which, in comparison with the traditional gonadectomy, has the advantages of easy administration (injection or implantation instead of surgical removal of tissue), reversibility, and controlled duration of effect. The most commonly used antihormones are the antiandrogens (for example, cyproteron and cyproteronacetate); female sex hormone can be inhibited with antiestrogens (for example, Tamoxifen).

Anting   Behavior observed in many birds, consisting of picking up ants and depositing them among the body feathers or on the skin. Related to anting, and often discussed with it, is similar behavior with nettles, live coals, as well as "bathing" in the embers of fires or hot volcanic ash. Numerous possible reasons for anting have been suggested in the literature, but opinion remains divided. The following suggestions have the greatest plausibility: the caustic exudate of the ants serves as a disinfectant; the formic acid helps to get rid of ectoparasites, such as mites; the irritants produce an exquisite sensation on the skin. The legend of the phoenix may have its origin in observations of fire-bathing birds.

Antiphonal singing   A song pattern in which two individuals sing parts alternately. It has been recorded in crickets, birds, and some mammals, such as squirrels and gibbons. (Compare DUET SINGING.)

Aposematism   See Warning behavior

Apotreptic behavior Threat behavior. In an effort to avoid the everyday connotations of such terms as "threat," "intimidation," and "attraction" in descriptions of animal social interaction, Barnett (1981) observed that most such interaction can be classified more neutrally and objectively in terms of approach and withdrawal (see APPROACH-WITHDRAWAL THEORY). He proposed that these social interactions be called treptics, from which he derived epitreptic behavior for action tending to cause a conspecific to approach and apotreptic for action tending to cause a conspecific to withdraw. These terms have not caught on.

Appeasement behavior A general term for patterns of behavior that can inhibit intraspecific attack or aggressive behavior by means other than intimidation. It occurs in social animals as greeting when two members of the same group or the two members of a pair meet, enabling them to approach more closely than would otherwise be possible. Appeasement often incorporates components of juvenile behavior, such as begging postures and movements, or of sexual behavior, as in many primate species where the female sexual presentation posture serves as an appeasement display. Darwin pointed out that what are now called appeasement displays are often the inverse of threat displays, exemplifying his principle of antithesis, according to which signals that are opposites in meaning are opposites in form. Thus an appeasing dog lowers its chest to the ground and points its nose upward, with ears back and flattened against the neck, while a threatening dog stands with chest as high as possible, nose pointed straight ahead, and ears erect and turned forward.

A distinction is sometimes drawn between appeasement and submissive behavior; according to this distinction, appeasement activates behavioral tendencies incompatible with aggression, and submission involves the cessation of aggressive signaling. More commonly, however, the term "appeasement" includes the connotation of submission.

Appetitive behavior Searching behavior; goal-directed behavior. Behavior that is varied according to the circumstances and that allows an animal to perform a specific kind of action or to attain a specific state of being, for example, satiety, reception of stimulation from brooding, arrival at a place suitable for resting, or contact with particular individuals. In the latter case appetitive behavior can be an expression of social attachment.

In its simpler forms appetitive behavior can consist merely of orientation and adjustment movements that are more or less reflexive in nature (see TAXIS). As a rule, however, it comprises a comparatively plastic series of different actions, which may include learned performances, such as tool use, as well as fixed action patterns.

In earlier ethological motivation theory, appetitive behavior was thought to be always directed toward finding the stimulus for a consummatory act (see END ACT), and the performance of the behavior was supposed to have no influence on the readiness for further action of the same sort. Since then, numerous cases have come to light of appetitive behavior directed toward consummatory situations and of its performance having feedback effects on motivational state or arousal independently of attaining the consummatory goal.

Besides these directed forms of appetitive behavior, there are cases of undirected appetence, which consist simply of increased locomotor activity, such as the random running about of rodents released into an open field. However, such activity can contribute to attainment of a goal by increasing the chances of encounter with the required object.

Applied ethology   A branch of behavior study concerned with animal species that are of direct practical interest to people, emphasizing the possibilities for practical application that are within ethology's domain. As most of the species with which people deal directly are either domestic animals or zoo animals, applied ethology more or less divides between veterinary ethology and zoo biology.

Approach-withdrawal theory   A view of behavioral development put forward by Schneirla (1965). According to the theory, when animals begin life they are capable only of forced movements; intense stimulation causes withdrawal, and mild stimulation causes approach. By interacting with its environment, the organism builds on this base qualitatively distinct levels of behavioral capacity, such as discrimination, motivational control, and hypothesis formation, which can be distinguished in comparisons of different kinds of animals as well as in stages of individual behavioral development. The theory is also referred to as a theory of biphasic processes.

Arena display   See Lek

Arousal   This term has a variety of applications in physiology and behavioral science. Roughly speaking, it is an animal's general state of excitability or activation. More specifically it can refer to the transition from sleep to wakefulness; the level of responsiveness, as indicated by the intensity of stimulation necessary to elicit reaction (see THRESHOLD CHANGE); the level of activation, as indicated by the kind of behavior exhibited (for example, relaxed grooming reflects a lower level of arousal than frantic fleeing); physiological indicators such as heart rate and skin conductance, as measured in lie-detector tests; and attentiveness to sensory stimulation. Not surprisingly, such multiple meanings have given rise to confusion and controversy from time to time, for the different indices do not always agree and so cannot be taken as alternative reflections of a single underlying variable. For example, heart rate remains high after an animal has exerted itself to the point of physical fatigue and after other measures of arousal are depressed. However, the relationships among the variables falling under the rubric of arousal are open to experimental investigation and present an important empirical matter for students concerned about the physiological bases and motivation of animal behavior.

An interesting aspect is the variation in patterns of electrical activity in the brain associated with changes in arousal, however measured. The vertebrate brain has a central core, the central gray or reticular formation area, which receives input indiscriminately from all the sensory systems and projects the pooled activation of this input, together with probable inherent excitation flux, as a background of neural facilitation to the more specific areas of the cerebrum, where connections are made between stimuli and responses. During sleep this reticular activating system is at its lowest ebb (although even in sleep there are cycles of "deep" and "shallow" levels, as reflected in arousability). During intense fear or hostile provocation the system puts out its maximum bombardment to the cerebral control areas. Arousal level is reflected in the pattern of electrical activity that can be picked up by electrodes from the surface of the brain as an electroencephalogram (EEG): when arousal is low, as in drowsiness or sleep, the neurons of the cortex tend to fall into synchronous rhythm, producing slow, high-amplitude EEG waves; when arousal is high the neurons "break step" as they get caught up in the separate circuits underlying, for example, perception, thought, and action, giving fast, low-amplitude (desynchronized) EEG waves.

Artificial selection   See Selective breeding

Association   The forming of learned connections between stimuli and responses or between stimuli and stimuli (see CONDITIONING).

The term is also sometimes used in a very different sense to refer to animals living together, either conspecifics or individuals of different species. In ecological contexts an association is a collection of species constituting a major unit in the structure of a community.

Assortative mating   Sometimes referred to as assortative pair formation. For many of the animal species in which it has been investigated, pair formation within a species or a population is not arbitrary; preference for specific phenotypes affects the individual's choice of partner (see SEXUAL SELECTION). As a rule the pairing choice is positive, which is to say that individuals choose one another on the basis of similarity of appearance in at least some features (positive assortative mating). To the extent that appearance mirrors underlying genotype, choice of a similar phenotype ensures that the pair will have much in common genetically. The biological significance of positive assortative mating probably lies in the fact that by combining similar genotypes each individual increases the representation of its genotype in the offspring (see FITNESS; KIN RECOGNITION; KIN SELECTION THEORY). Also, local adaptation—genetic adaptation to the conditions of the current habitat—can in this way be better promoted and maintained. On the other hand, pairing between closely related individuals can lead to the deleterious effects of inbreeding. In many species this is prevented by some form of incest avoidance. The usual compromise, as sociobiologists have shown, is that natural selection promotes pairing between related, but not too closely related, individuals; this is the theory of optimal discrepancy. In forming the preferences leading to assortative mating, learning may play a role, especially when young are still living in the family group (compare SEXUAL IMPRINTING, which can contribute to reproductive isolation).

Atavism Throwback inheritance; reversion to type. The occasional occurrence, in individuals of a species, of characteristics of phylogenetically ancestral forms. Atavism is distinguished from relics and recapitulation in two regards: it occurs as a rule only sporadically in a population, and the characteristics may periodically disappear— sometimes for many generations—then suddenly recur.

Atavism is best known from morphological cases, such as the occasional formation of a second or third hoof on the legs of perissodactyls (horses and their kin) and a second pair of wings in dipterous (two-winged) flies. It arises from mutation, from disturbance during embryological development, and sometimes from hybridization between members of closely related species. Hybridization results in behavioral instances of atavism. Many interspecies hybrids among ducks and among parrots lack the behavior patterns of the parent species and in their place show patterns of the common ancestral species from which the parent species derived.

Attachment A tie between an animal and some place, object, or social companion. A nonsocial attachment can be to a retreat, such as a cave or a hole in a tree, to a nest site or breeding territory, or to a specific habitat or locality to which the animal returns at certain seasons year after year, as in the case of migratory birds (see FIDELITY TO PLACE). Place or habitat attachments are often consequences of early experience (see PLACE IMPRINTING). For social attachment, see BOND.

Attention The direction of an animal's interest or concern at any moment. An animal cannot attend equally to all of the stimulation its sensory systems are subject to at any one time; it must selectively attend to whatever is salient. Such salience can result from the filtering of stimuli; from inhibition of one system or part by another; from the tuning out of inconsequential stimuli through habituation; from arousal caused, for example, by a sudden encounter with novel stimulation (orienting response). Shift of attention was demonstrated neurophysiologically by an experiment in which an electrical recording of the auditory system of a cat showed evoked potential volleys (see NERVE IMPULSE) corresponding to the clicks of a metronome until a mouse was let loose in the cat's vicinity; the response to the auditory stimulation promptly ceased as the cat focused on the prey. A form of selective attention familiar to ethologists is that of foraging animals governed by a search image: the animals attend to the most abundant or discernible food and may fail to notice equally acceptable food that could become the focus of attention if the other source becomes depleted.

Attention structure   The directions of regard among animals in a group in relation to their social status and spatial position. It is part of a view of social organization that is reminiscent of topological field theory in social psychology. Each individual is said to occupy a social space or field structured by its relationships and the tensions it is subject to within the group. The structure of the group as a whole is manifest in who pays attention to whom and how the individuals move with respect to one another. This approach to social organization has been applied mainly to primates, in which it has been found to take different forms in different species. Attention structure has also been applied to the interpretation of some human situations.

Autecology   See Ecology

Autism   A condition encountered in human psychology, in which individuals who are shut off to varying extents from social contact with other people show such symptoms as persistent stereotyped movement and reduction of sensory responsiveness. Captive animals sometimes display a similar syndrome, especially where the conditions of captivity lack features needed for normal behavioral development (see DEPRIVATION SYNDROME).

Autochthonous drive   See Allochthonous drive

Autonomic nervous system   The division of the vertebrate central nervous system that innervates and hence mediates control over the viscera, including the alimentary canal, heart, blood vessels, and certain glands. It consists of two parts that are antagonistic to one another in their effects. The sympathetic division takes control in

23

situations of emergency or stress, mobilizing energy reserves, suppressing digestive processes, increasing heart rate and flow of blood to the muscles and brain, deepening breathing, and causing other changes conducive to performing actions or withstanding trial in extremis. The parasympathetic system reverses the effects of the sympathetic system, returning heart rate and blood supply to normal, relaxing the curbs on digestive processes, and so forth.

The physiological states brought into being by autonomic activity can manifest overtly as, for example, in urination and defecation, feather ruffling, or hair erection. In many cases such symptomatic expressions have become ritualized, with the addition of morphological support, and hence have become specifically communicative in function.

Autoshaping When animals are exposed to certain stimulus contingencies, for example when a light is turned on and food is delivered, they may, after a number of such pairings, start behaving toward the light as though it were food. A pigeon treated in this way will develop the habit of pecking at a lighted key without being trained specifically to do so in the usual manner of operant conditioning. Similarly, if the stimulus becomes associated with opportunity for sexual activity the bird will begin acting sexually toward it. The association of the stimulus and the reward apparently conditions the animal in the classical manner, the transferred behavior being comparable to the anticipatory salivation of Pavlov's dogs.

Autumn song See Subsong

Aversion-induced aggression Pain-induced aggression. Attack behavior consequent on aversive stimulation. It may be directed at any bystander or inanimate object that is within reach. The term comes from psychology, where the phenomenon has been studied experimentally, but in ethological terminology it is comparable to redirection (see also FRUSTRATION-INDUCED AGGRESSION).

Avoidance behavior See Avoidance conditioning

Avoidance conditioning That which leads to conditioned avoidance. Stimuli that are initially neutral or even sought-after come to

24

be avoided by an animal as a consequence of an associated aversive experience (fright, pain, nausea). The avoidance behavior can consist of simply staying away from a place or object (passive avoidance), refusal to approach or contact it, or active flight from it. It is common in connection with food selection, especially in eutrophic animals such as rats, whose diet is extremely varied and variable.

Awareness   See Consciousness

Axon   See Nerve impulse

Azimuth   Horizontal direction with respect to some reference point. In ethology the term is used mainly in connection with navigation; the reference point is usually the position of the sun. In other fields, such as astronomy, azimuth can mean the clockwise angle between a particular horizontal direction and due south.

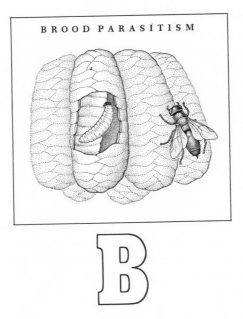

BROOD PARASITISM

B

Badge of status   A term introduced by Dawkins and Krebs (1978) for features of appearance that indicate ability to maintain control of a resource, such as territory (see RESOURCE-HOLDING POTENTIAL). For example, the dominance status of male Harris's sparrows (*Zonotrichia querula*) correlates with the area of black feathers below the beak. When the size of this bib was experimentally enlarged on some subordinate males, higher-ranking males attacked them more, showing that the wearers of badges may be tested to see whether the advertised status is merited.

Basic rank   The paired terms "basic rank" and "dependent rank" are used mainly in the primate literature to signify that the positions held by two individuals in a rank order may differ according to the presence or absence of other social influences. Basic rank is the rank relationship between two animals in situations where other group members do not interfere; dependent rank is their rank status as influenced by other group members or the group as a whole. The two rankings may be quite different, yet they may also be connected, in that the longer an animal has been at a higher dependent rank the higher its basic rank is likely to be (see CENTRAL HIERARCHY).

The best-known example of dependent rank is that of the young in many primate groups: though their basic rank is very low in one-on-one encounters with adults, their status approaches that of the mother within the group. Primatologists have discussed three possible mechanisms for the existence of this situation: it may be genetically based; or the young may be identified with the mother through, for example, imitation of her facial expressions and gestures (so-called rank mimicry); or the mother may actively support the young in altercations with other young. Of these possibilities the first is the least likely and the last the most likely.

Dependent ranking has also been observed in species that have separate rank orders within each sex. Thus, for example, in jackdaws the rank of a female corresponds with that of the male she is paired with.

Batesian mimicry   See Mimicry

Beau Geste hypothesis   See Song repertoire

Bee language   See Language

Behavior   What ethologists study in animals. However, one ethologist's description of an animal's behavior in a particular instance may differ from what another would describe. There are usually alternative ways of differentiating and classifying what animals are doing, and these alternatives may not be collapsible into one another. One kind of description is in terms of change of position of parts of the body relative to other parts and to environmental coordinates. This includes movement categories such as extension and flexion at one level of detail, running and flying at another level. A different kind of description is in terms of consequences or the outcome aimed at; it categorizes the behavior as an action such as attacking, fleeing, threatening, or nest-building, sometimes implying intentionality but more often, in ethological contexts, referring only to functional consequences. It is an ethological precept that description in terms of motor pattern and description in terms of consequence or functional categories should not be confused.

Behaviorists, of course, also study animal behavior. They recognize a distinction between molecular behavior, corresponding to movements, and molar behavior, corresponding to actions.

Behavioral characteristic   The usual meaning of characteristics in biology is "all realiably recognizable attributes or distinguishing marks of an organism" (Wickler and Seibt 1977). The term "behavioral characteristic" refers to any such feature connected with behavior, for example a movement pattern, vocalization, scent signal, sensitive period, or responsiveness to a sign stimulus. The term is useful because only a few of these characteristics are covered by such general terms as "behavior pattern" or "action pattern."

Behavioral ecology   Ecoethology, sociobiology. These are new designations for a part of ethology that is as old as the science itself: the study of the relationships between the behavior of an animal species and the physical and biological conditions under which it lives. This branch of ethology has been especially concerned with parallel and convergent behavioral adaptations: cases of species not closely related to one another that have independently evolved similar ways of coping behaviorally with habitats presenting similar selection pressures (habitat-typical behavior). Convergent adaptation is most evident in foraging and reproductive patterns. For example, flower-exploiting insects of different orders (such as Lepidoptera, Hymenoptera, Diptera) have independently evolved the ability to hover in place, enabling them to "stand" in the air in front of the flowers from which they draw nectar. Hummingbirds provide an example of vertebrate convergence on this adaptation.

Among other aspects of behavior showing adaptation to the environment, variations in social systems have perhaps attracted the most attention as subject to adaptive correlation. For example, forest-dwelling species of tropical and subtropical birds live predominantly in pairs, while those that live in open habitats (savannah, steppe) usually live in groups. Among other things, this difference correlates adaptively with the nature of the food supply. In the forest the food is temporally and spatially quite evenly distributed and is therefore more efficiently exploited by equally spaced, uniformly distributed sedentary pairs. In open and dry habitats the food supply is patchy, being abundant in one locality today, in another locality tomorrow.

Such a transient and uneven food supply is best utilized by groups, whose members can inform and lead one another to food sources. This is a relatively simple example of the kind of relationship between behavior and ecology, social structure and environmental circumstance that behavioral ecologists and sociobiologists seek to account for by means of theoretical models such as optimal foraging theory (see OPTIMIZATION THEORY).

Behavioral embryology   The branch of ethology that is concerned with behavioral development before birth or hatching from the egg. One of its important discoveries, which applies to many different species, is that in embryological development coordinated motor patterns appear before the related sensory receptors are functional, that is, before external or peripheral stimulation can exert an influence on the central nervous system. Behavioral embryology has thus contributed to the evidence for the partly spontaneous nature of behavior and has provided detailed analyses of environmental and hereditary components in the genesis of behavior. Also it has been able to show that in many birds learning can occur while a chick is still in the egg (see PRENATAL LEARNING).

Because the embryological stages present relatively simple and easily comprehended causal relationships, behavioral embryology has extensively traced the joint development of behavior and the central nervous system, in attempts to elucidate the controlling functions of the latter. This approach is sometimes referred to as the neuroembryology of behavior.

Behavioral endocrinology   That part of behavioral science concerned with the influence of hormones on behavior, and of behavior on hormonal secretion. It is sometimes called ethoendocrinology.

Behavioral estrus   See Estrus

Behavioral genetics   Investigation of the hereditary transmission of behavioral traits and of the ways genetic factors work together in affecting behavior. Its methods are mainly those of classical genetics, such as cross-breeding and artificial selection. By these means ethologists and others have demonstrated that behavior patterns and dispositions can be passed from one generation to another according to the same principles of heredity that apply to bodily characteristics and functions. Another part of behavioral genetics, which is more akin to genetics itself, examines very simple, easily comprehended behavior patterns to pursue genetic questions, such as the difficult one of how single genetic factors influence specific behavioral elements. The behavioral consequences of excluding single genes are compared with the corresponding behavior of untreated animals. This approach has been directed almost totally to work on fruitflies (*Drosophila*), the only group for which genetic knowledge is sufficient to allow this kind of investigation.

Behavioral inventory   See Ethogram

Behavioral mimicry   Ethomimicry. A species' imitations of behavioral rather than morphological characteristics of other animal species. Thus carnivorous fish mimic the conspicuous swimming movements of certain cleaner fish, enabling them to approach the "clients" as though to pick off parasites (see CLEANING SYMBIOSIS) and bite off small pieces from the fins. These fish mimics are also shaped and colored like their models, the behavioral imitation thus having morphological support. Among the numerous alien species inhabiting ants' nests, called "ant guests" (beetles, silverfish), some mimic the antenna movements of the reciprocally feeding ants and so elicit regurgitation of food from their ant hosts. Occasionally such mimicry occurs through vocalization. Thus young burrowing owls, faced with a threat from a predator, make a call that is almost identical to the sound a rattlesnake makes with its tail, which can frighten the predator away.

Behavioral ontogeny   See Ontogenesis

Behavioral phylogeny   The branch of ethology concerned with the evolutionary origin and descent of behavior patterns. Because of the almost total lack of fossil evidence, behavioral phylogeny is in a very difficult situation methodologically. Statements about extinct animal forms, which have contributed so substantially to the reconstruction of the course of evolutionary change in the morphological domain, are possible only to an extremely limited extent with regard to behavior. Inferences about the phylogenetic history of behavior patterns must be drawn almost entirely from observations of the behavior of currently living species. The two possibilities are study of behavioral ontogeny and comparison of closely related species whose taxonomic position and supposed evolutionary level within the taxonomic group are already well established through systematics and other parts of zoology (compare COMPARATIVE ETHOLOGY).

Behavioral physiology   That part of ethology that deals with the physiology of behavior, and that part of physiology that deals with behavior. It includes neuroethology and behavioral endocrinology.

Behavioral program   See Program

Behavioral repertoire   The totality of behavior patterns characteristic of a species. Ethologists may disagree about what constitutes a particular repertoire because they categorize the behavior in different ways (compare ETHOGRAM).

Behavioral science   A collective term for all empirically based investigations concerned with behavior, including ethology and comparative psychology (behavior of animals); human psychology, human ethology, and various divisions of sociology and anthropology (behavior of humans); and behavioral physiology.

Behavioral system   See Functional system

Behavioral tendency   See Action potential

Behavioral vestige   See Relict

Behaviorism   A school of psychology founded by the American J. B. Watson (1913) and having its most powerful influence in the United States. Its basic and revolutionary premise was that psychology should be regarded as the science of behavior as opposed to the science of mental life, as William James believed. This implied that all mentalistic concepts were to be eschewed, unless they could be redefined in terms of observable behavior and the ascertainable or contrived circumstances of its occurrence. All psychological concepts must be reducible to those terms. There is considerable variety among the positions of different behaviorists, but they share a bias toward environmentalism, the belief that all behavior apart from the simplest reflexes is a product of learning. Although there are exceptions, behaviorists tend to be atomistic, believing that behavior is a mosaic of basic reflex units; physicalistic, believing that behavior must ultimately be explained in terms of physics and chemistry; sensationistic, believing that all behavior is response to stimulation and hence never truly spontaneous; and universalistic, believing in general laws of learning. This last is reflected in the usual assumption that one can choose the kind of animal and the learning situation to be studied on the basis of convenience, without regard to differences between kinds of animals or kinds of learning situations. Hence most of the research has been done on a very few types of animals, such as white laboratory rats and domestic pigeons, and in a few simple and standardized laboratory learning situations, such as the maze and the Skinner box. Ethology has been critical of behaviorism on many of these points.

Cognitive psychology and cognitive ethology have attacked the behaviorists' prohibition on mentalistic concepts, arguing that many of the central questions of psychology and even ethology can be framed only in terms of such concepts. Behaviorism has undergone decline as the cognitive and biological approaches to mind and behavior have gained ground.

Billing   Bill-to-bill nibbling in the courtship of birds. It often includes feeding and food-begging movements and even actual transfer of food, from which it is no doubt derived (see COURTSHIP FEEDING).

Bioacoustics   A branch of zoology that investigates sound production and reception in animals. Its development dates effectively from about 1950, when practical precision-recording methods and equipment became readily available. Perhaps equally important was the invention of the sound spectrograph and its kin (see SOUND SPECTROGRAM), which made possible the objective analysis and graphic reproduction of animal sounds, thus dispensing with the very subjective transcriptions in words and musical notation that many ornithologists had used. Bioacoustics also includes study of the organs of hearing and the sound-generating apparatus (vocal

32

cords, stridulation mechanisms, vibrating membranes and so on) as well as the study of the physiological processes by which sounds are produced and received. Finally it attempts to find relationships between the characteristics of the sounds an animal produces and the nature of the environment in which they are used and the functions they are designed to serve.

Bioacoustics has yielded important ethological information, especially in two areas: the ontogenetic development of vocalization and the relative contributions of genetic inheritance and experience to it, and acoustic communication within and between species.

Biocommunication See Communication

Biogenetic law See Recapitulation

Biogenic social regulation The control of behavior in social groups by biological factors such as nutritional need, temperature, and humidity conditions, together with the signal mechanisms conveying information about such factors between the members of a group. The term was introduced by Schneirla (1949) to mark a difference between the ways insect societies are regulated and the ways the social behavior of higher vertebrates is regulated (see PSYCHOGENIC SOCIAL REGULATION). It is now little used.

Biological rhythms Biorhythmicity. The periodic phenomena (daily, monthly, and annual cycles) of life and their underlying physiological regulation. Since such periodicity is expressed very prominently in behavior, the study of biological rhythms has an important place in ethology.

Biotope Habitat. An autonomous region, distinguishable from other such regions by the living conditions it presents and the associations of plants and animals adapted to it. Biotopes are usually labeled by the predominant vegetation type, such as sprucewood, moorland, steppe. The organisms characteristic of a biotope constitute what is described as an ecological community or biocenose.

Biphasic processes See Approach-withdrawal theory

Bond Social attachment; a specific dependence relationship between two or more individuals. The most common bonds are between the partners in mated pairs (see PAIR BOND), between young animals and their parents or the mother, and among members of groups, especially individualized groups (see GROUP COHESION).

In particular instances it can be very difficult to determine whether bonding really exists. The criterion of spatial proximity, which is frequently used in human psychology, has only limited application in ethology, for numerous animal gatherings in which the individuals are close to one another, even touching, lack true bonding. Ethologists must resort to other criteria, for example a certain exclusiveness of relationship between individuals. Thus we speak of bonding when one animal behaves toward another, or a limited number of others, in special ways, such as tolerance in situations of discord, or with particular patterns, such as presentation among primates. Sometimes the frequency of certain actions can be a mark of a bond between two individuals. In some cases a distinct motivation, referred to as a bonding drive or social drive, is believed to contribute to the forming and maintenance of a bond. Compare ATTACHMENT.

Bonding behavior A general term for all behavior that serves to establish or maintain a bond. The bond between the partners of a pair is usually formed and sustained with the aid of sexual behavior patterns, many of which are identical to precopulation courtship. Other behavior that promotes bonding includes duet singing, courtship feeding, and allogrooming. Sexual patterns can also play a role in group cohesion, sometimes with components from other kinds of behavior, such as parental care. In some cases, especially where bonding persists year round and so extends beyond the breeding season, the bonding behavior appears to be governed by its own motivation.

Bonding drive Motivation that draws the members of social species together. The concept is disputed, in part because drive concepts in general are problematic and because even those ethologists who hold to a drive concept are divided about whether there is evidence for a drive for social attachment independent of other kinds of motivation, such as those associated with sexual activity and parental care. Two cases of pair bonding have been claimed as definitely implying an autonymous bonding drive: the triumph ceremony of greylag geese and the pair sitting of Indian and Pacific Ocean prawn species. In both cases the behavior patterns have their own appetitive behavior. Other observations of group formation, especially in insects, fish, birds, and mammals, have also been interpreted as implying that the social union is governed by its own motivation, distinguishable from other tendencies. Alternative terms in the literature include social drive, herd instinct (of mammals), and, to avoid the controversial connotations of drive, the more neutral-sounding social tendency and gregarious disposition.

Bower   See Display territory

Brain   See Central nervous system

Brain stimulation   The stimulation of the brain with weak electrical current, usually by means of microelectrodes, insulated except for the tip, which are inserted into specific brain loci. Electrical current can elicit behavior patterns or change motivational state and thereby help to demonstrate the roles of particular brain structures in the regulation of behavior. Most brain stimulation studies have been done on vertebrates, mainly mammals and birds, but a little work has been done on invertebrates also, mainly insects and crustaceans. The techniques of brain stimulation are among the main methods of neuroethology.

Breeding colony   This term is applied mainly to nesting concentrations of communally breeding birds, among which are some of the most striking examples of coloniality. Colonial breeding can be induced by extrinsic factors such as restriction of adequate nest sites (as for many oceanic birds), but it can also have an intrinsic basis in forces of social attraction. Among many weaver finches, waxbills, and starlings, several or many pairs nest together even though other, equally suitable nest sites are available in the region. The influence of restricted resources suggests that the colony can be adequately described as an aggregation, yet as a rule the level of social interaction, at least between neighbors, is sufficient for most breeding colonies to be regarded as instances of group formation. Benefits of coloniality include better predator protection through warning and collective defense and, possibly, the effects of reciprocal stimulation.

Broken-wing display   See Distraction display

Brooding   In birds, the parent's sitting, squatting, or crouching over its brood of chicks to provide warmth and cover. Sometimes brooding has the broader sense of being engaged with the care of the brood, even extending to the incubation of the clutch. For precision, the term should be reserved for the covering of chicks, and incubation for the covering of eggs.

Brood parasitism   A form of parasitism in which the parasitic species partly or wholly leaves brood provisioning or parental care to members of the host species. Brood parasitism requires special adaptations for the deposition of eggs or young, and such adaptation can extend to significant portions of the behavior of both adults and young. It is therefore by no means a more primitive or simpler form of parental care. The best-known example of a brood parasite is the cuckoo. Cowbirds and widow birds (Viduinae) are other avian examples, and various insects, mainly beetles and Hymenoptera (cuckoo bees), leave their larvae to be raised by members of closely related species.

Brood patches   Incubation patches. In birds they are areas of ventral skin that become defeathered, vascularized, and edematous during egg laying and stay that way throughout incubation. When incubating, a bird applies its brood patch or patches to the eggs by sitting on them, thus facilitating transfer of heat from its body to the eggs. The formation of brood patches is hormonally induced. Some species form one large patch; gulls have three, which is the usual number of eggs in the clutch. In species such as songbirds, in which only one sex incubates, it alone has brood patches; where both sexes incubate, both have brood patches as a rule. Ducks, pigeons, gannets, and a number of other birds lack brood patches. Gannets and their kind apply the inner webs of the feet to the surface of the single egg when the birds sit to incubate, but it is not known whether heat transfer is enhanced by physiological changes analogous to those of brood patches.

36

Brood provisioning A general term for all behavior of parent animals anticipating the arrival of young by providing conditions to promote their development: the establishment and maintenance of sheltering burrows, nests, or cocoons; the gathering of a food supply or the laying of the eggs in the vicinity of a food source sufficient for the hatchlings' needs. Generally the term covers only the period up to the deposition of eggs. Where parental concern persists beyond that point and leads to direct contact between adult and young, we usually speak of parental care.

Bruce effect In some strains of mice, exposure of a pregnant female to odor from a male other than the one that fertilized her can cause the developing fetuses to abort. The effect is named after the person who discovered it.

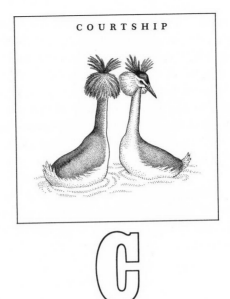

COURTSHIP

C

Cainism   See Fratricide

Call   In the ethological, ornithological, and bioacoustical literature, a vocalization or voiced sound that is much simpler than the songs of songbirds. Calls differ from songs in their brevity and scanty syllabic structure. However, the borderline is difficult to draw, for on the one hand some call sequences are lengthy, and on the other many songs consist of only a few elements or even a single repeated syllable type. For birds function can be a means of differentiation, for song generally is used only to advertise territory or attract a mate, while calls are used in a variety of other behavioral contexts. In other animal groups (insects, amphibians), however, calls or call sequences can serve the territorial and mate attraction functions of birdsong. For example, in the two-syllable call of the tree frogs of Puerto Rico, one syllable serves to attract females, while the other syllable keeps males at a distance. A similar double function is supposed for the calls of other frog species also.

As a rule different functions go with different calls, so we speak of luring, contact, nest-soliciting, leading, warning, and begging calls.

Camouflage Crypsis; disguise for concealment. Many animals have surface coloration or markings, shapes, or postures that make them difficult to discern against the background of their normal habitat. A common form of camouflage is countershading: the upper part of the body is darker than the lower, the reverse of the gradient of reflection when light falls from above. The effect is to eliminate or reduce the three-dimensional modeling that light would give to the body, which consequently looks flat when viewed from the side. Thus fish are typically dark dorsally and light ventrally. An exception is the upside-down catfish (*Synodontis*), which has a dark belly and light back (reverse countershading); this is an exception that proves the rule because the fish normally swims on its back. Similarly, many caterpillars are reverse countershaded and move with their ventral surface toward the light. One form of disruptive coloration consists of lines or stripes that cut across the contours of the body, thus breaking up the appearance of its form, as in tigers and zebras. The eyes, a particularly vulnerable part of the body, are often concealed within a stripe or by a stripe running through them. Parallel stripes or bands can also serve to merge a body with its background by conforming to linear features such as grass blades or reed stems. For this to work, the animal may have to adopt a specific posture, as is the case with the bittern, whose longitudinal striping blends with the reeds when the bird stretches its body skyward. Some fish and crustaceans achieve camouflage through transparency. Another common form of camouflage is mimesis: disguise to resemble an object that is likely to be ignored or overlooked by the audience it is designed for; examples are stick insects, walking leaf katydids, and thorn-shaped tree hoppers.

Camouflage features can be either fixed, like a leopard's spots or an egg's speckles, or changeable, like the mottling of a flounder or the color of a chameleon, which alters with the animal's surroundings. Arctic hares and ptarmigan alternate seasonally between snowy white and the color of tundra straw. Cephalopods, such as the octopus and cuttlefish, are protean quick-change artists. Some animals switch between concealing and conspicuous appearances, as in the case of moths that can flash an eyespot display when their camouflage has been penetrated (defensive behavior), and species such as fiddler crabs, in which crypsis alternates with episodes of flashy social display. Where there is reproductive division of labor, as in eider ducks, the sex saddled with parental care is often camouflaged and the other is conspicuously adorned.

Camouflage serves for protection or defense against predation, and also for predation itself, enabling a predator to lurk in ambush for passing prey, or, along with Peckhamian mimicry, to lure prey within striking reach, as anglerfish and snapping turtles do.

Cannibalism Feeding on conspecifics. Cannibalism has been described in a large number of animal species, from protozoans to primates. It occurs as much in herbivores as in carnivorous species. In some spiders and insects such as the praying mantis, ground beetles (Carabidae), and empid flies, the female eats the male after copulation. In some cases an association has been found between cannibalism and food shortage. Here the consuming of conspecifics obviously contributes to the cannibals' nutritional needs. In other cases the practice occurs even when food is in abundance. Cannibalism then probably contributes to regulation of population density, working against the whole population's being jeopardized by overcrowding. Special forms of cannibalism are killing and eating one's own young (infanticide, cronism), and killing and eating of siblings (fratricide or siblicide).

Carrying in Transport of objects (for example, nest material or winter provender) or young to the nest, den, or burrow. The carrying in of young is especially common among rodents and carnivores; when they perceive that a pup is outside the nest, they use their teeth to pick it up and carry it carefully back into the nest. In this manner the whole litter can be moved to another nest in the event of disturbance. While being carried, the young tend to hang limply in the parent's jaws. This retrieval of the young is noteworthy in that it is one of the few behavior patterns that virtually never wane in responsiveness (see ACTION-SPECIFIC EXHAUSTIBILITY, HABITUATION), thus ensuring that all pups in a litter get transported during a nest transfer, even when the distance to the new nest is very great.

Similar retrieval of young has been observed in some birds. Analagous patterns are the collecting of nest material and retrieval of
eggs by birds; the retrieval by fish of eggs (dwarf gourami) and of
fry (African mouthbreeder); and various types of hoarding, provisioning, and retrieving by social insects.

Caste In biology the term is applied to the social insects. It refers
to a class of individuals specialized in behavior and morphology to
carry out specific functions. In contrast, in the majority of animal
species, individuals have a variety of roles, and morphological variation, if any, is restricted to that between the sexes (sexual dimorphism). Among the social insects the commonest castes are queens,
the only females capable of producing offspring; soldiers, which
defend the colony; and workers, which forage, build, maintain, and
take care of the brood. (See DIVISION OF LABOR.)

Castration See Gonadectomy

Cell assembly In an influential theory of the neurophysiological
basis of learning and memory, Hebb (1949) proposed that the basic
unit of storage in the brain is a group of neurons connected in a
ring or "closed circuit," which he called a cell assembly. Initial learning or short-term memory involves the temporary coursing of excitation through such a circuit. When repeated sufficiently often, this
coursing causes lasting synaptic changes that constitute the engram.
In a similar way a number of cell assemblies can be connected in a
phase sequence to encode more information.

Central hierarchy In certain primate species in which rank order
governs interactions within groups, alliances among males can affect
the individuals' dominance standing. The members of the alliance
can call on one another for support in disputes and thus together
dominate the troop, even though some excluded individuals might
be able to defeat each member in single combat. The members of
the alliance are not likely to be on an equal footing when they must
compete with one another, so the alliance will have an internal rank
order. An alliance of this sort is referred to as a central hierarchy.
Central hierarchies have been observed in the social organization of

cynocephalus baboons, rhesus macaques, and a number of species of New World monkeys.

Central nervous system   CNS for short. That part of the nervous system lying interior to the skin and external musculature (hence central), which contains the majority of the cell bodies of the neurons and the synapses between neurons. The CNS is the seat of the highest functional achievements of animals. It coordinates the activity of individual organs, processes the incoming messages from sense organs, and, in accordance with the sensory and stored information, regulates the behavior of the animal as a whole. It is the locus of learning and memory. The most highly evolved central nervous systems are those of vertebrates, which consist of a brain and hollow dorsal spinal cord, and those of arthropods, which consist of hypertrophied supra- and subesophageal ganglia and a solid, double ventral nerve cord with ganglia in each segment.

The term "brain" usually refers to the vertebrate structure, which dominates the rest of the central nervous system, but the term is also sometimes loosely used for anterior enlargements having an analagous dominating function, such as the head ganglia of arthropods and worms.

Ceremony   Also "ceremonial." The two terms were more common in the earlier literature of comparative ethology than now. They refer to complex social behavior sequences distinguished by a high degree of constancy of form, for example, courtship ceremony, greeting ceremony, nest relief ceremony. See also RITUALIZATION.

Change of function   Evolutionary change in the function of a bodily structure or behavior pattern. The classic example from comparative anatomy is in the vertebrate skull: from the bones that form the jaw articulation in fishes, amphibia, and reptiles, the mammals evolved the middle ear ossicles malleus (hammer) and incus (anvil), which are important parts of the auditory apparatus. Concurrently the function of jaw articulation passed to other skeletal elements (secondary jaw articulation of squamosal and articular bones). Behavioral examples are the components of food begging, nest building, and other activities that have become constituents of courtship behavior. Indeed the concept applies to the evolution of courtship in

42

general, according to the teaching that displays are derived activities, adapted to a communication function through ritualization. Compare EMANCIPATION.

Character displacement Where the ranges of distribution of closely related species with similar ecological niches overlap, the differences between them tend to be more pronounced than in those parts of their ranges that they have to themselves. Apparently, in the region of overlap, selection has favored the partitioning of resources through divergent specialization rather than head-to-head (species-to-species) competition which, according to the competitive exclusion principle, would end in the elimination of one of the species, at least in that region. Since interbreeding would tend to undo the effects of selective specialization and would produce hybrids less well adapted to exploit the resources than offspring of parents mating with their own kind, barriers to interbreeding are likely to develop (see REPRODUCTIVE ISOLATION). To counter the possibility of misalliances at pair formation, any means by which members of one species can be differentiated from the other, such as differences in appearance and in courtship behavior, tend to be exaggerated through selection. A well-analyzed example of this is the mate calling of two species of Australian frogs, *Hyla ewingi* and *Hyla verreauxi*: in areas where they are sympatric their calls are more distinct than they are where the two species are allopatric. (See also ISOLATING MECHANISM.)

Chemical castration See Gonadectomy

Chemoreceptor A sense cell that reacts to chemical stimulation. Taste and smell are the familiar modalities of chemoreception. Other chemoreceptors include brain cells whose function is to register the glucose concentration of the blood (see INTEROCEPTOR).

Chemotaxis See Taxis

Child schema The combination of bodily and facial features that arouses in people the tender sentiment they typically experience toward young children. The schema includes large roundish eyes, a large head in proportion to the body, small nose and chin, and rounded, chubby cheeks. These features characterize the human infant, but also appear in the juvenile characteristics of some other animal species, with similar effects on the feelings and actions of people who are affectionate toward children. The child schema has therefore been viewed as an example of an innate releasing mechanism in humans. However, without rigorous investigation of the contribution of experience to the instilling of these reaction tendencies, it is idle to speculate about a possible genetic basis for them.

In the toy and advertising industries the characteristics of the child schema are often presented in exaggerated form, for example in dolls and cartoons. Also many house pets, such as Pekinese dogs, are evidently favored, even as adults, because of their pronounced child features.

Choice test   A method for determining the social preferences of animals and for investigating their learning and discrimination capacities and the releasing and directing effects of stimuli. A choice test presents the experimental animal with two or more objects or live animals, for example on a selection platform or at opposite ends of a multipartitioned cage or aquarium. The animal's reactions or the time it spends in the vicinity of each option are recorded. As a rule the choice objects are presented together (simultaneous choice test), but sometimes they are presented one after the other (successive choice test). The results of simultaneous and successive tests do not always agree when given with the same alternatives, and the lack of agreement can have biological significance. For example, when gull chicks heard recordings of their parents' calls and comparable calls of neighboring adults, they recognized their parents' calls in simultaneous presentations, but showed no discrimination in successive presentations. In the natural situation chicks of this age are better off with their parents, but if they get separated from their parents, as can happen during flooding of the gullery, their only hope of survival is to be adopted into another family, so indiscriminate responsiveness is then the more adaptive course.

Chorusing   See Communal song

Circadian rhythm   Diurnal rhythm; an approximately twenty-four-hour rhythm. The daily cycles of behavior and other functions of the body. In many cases the rhythm is presumed to be endogenous, that is, it is derived from within the organism and is synchronized with environmental periodicity through a zeitgeber. Such synchronization is necessary because the endogenous periodicity is never exactly twenty-four hours, but rather a little longer or shorter. The period of the endogenous rhythm can be determined by subjecting the experimental animal to constant environmental conditions, such as continuous light of fixed intensity. Under these conditions the rhythm continues as a self-timed oscillation, which gradually gets out of phase with the external cyclicity that serves as the zeitgeber under normal conditions. Such oscillation in the absence of external influence is referred to as the free-running rhythm.

Circannual rhythm, circannual periodicity See Annual periodicity

Clade See Cladism

Cladism A school of taxonomy that bases its classification of organisms on evidence of common ancestry. Cladistic taxonomy is thus an exclusive hierarchy designed to match the strictly divergent branching patterns of phyletic descent. Its categories are called clades (from the Greek word for branch); a clade is a group of organisms all descended from, and the only descendants of, a single ancestor. Thus mammals and birds constitute clades; reptiles and amphibians do not, however, because the common ancestor from which each of these groups arose also gave rise to other groups. Cladists portray their taxonomies as branching diagrams called cladograms.

Although cladism is a persuasively logical approach to taxonomy and has attracted a large following in biology, it is opposed by approaches that place more weight on similarity between groups and on change within lineages than on phyletic branching. Some schools even eschew evolutionary considerations altogether.

Clasp reflex Grasp reflex. Reflex flexion of figures or toes and adductive gripping of arms or legs in response to touch on the palms of the hand or soles of the feet. The clasp reflex is best known from batrachians and arboreal mammals. However, it has quite different functions in these two groups. In frogs and toads the clasp reflex enables sexually active males to cling to the backs of females and fertilize the eggs as soon as they emerge from the female's cloaca. To this end the males in some species develop horny patches of skin, or nuptial pads, on their hands. The clasp or grasping reflex occurs in the young in species of tree-living mammals in which the mother carries the young (see CLINGING YOUNG). It consists of an ordered series of finger or toe flexions by means of which the infant can hold fast constantly to the mother's body. Clasp reflexes occur in newborn human infants also and are especially pronounced in premature babies. They present a good example of vestigial behavior.

Classical conditioning   See Conditioning

Classical ethology   See Ethology

Cleaning symbiosis   A form of association between animals of different species that benefits them both. The requirement of one partner is to be rid of ectoparasites; for the other the ectoparasites constitute an important or even the sole source of food.

In the literature the term is applied most frequently to fishes and decapod crustaceans, in which such associations are unusually common: more than forty species of fishes have been categorized as cleaner fish, and an even larger number of species of shrimps and prawns are known to gather ectoparasites from certain fish. As a rule the cleaner fish are conspicuously colored, probably so that their "clients," which may include predatory fish that treat other species of similar size to the cleaner as prey, can recognize them for what they are. Their coloration thus presents a good example of an interspecific releaser. The cleaner fish also show adaptations of behavior such as specific body positions and conspicuous ways of swimming. The hosts, similarly, may make behavioral adjustments, such as spreading the gill covers to enable the cleaners to gain access to the gills, where parasitic infection can be especially heavy.

Cleaning symbiosis occurs in other animal groups as well. Thus many birds pick parasites off the skin of large mammals or reptiles (crocodiles, tortoises, iguanas), and a similar relationship has been described between a mite and a beetle species, and between a rotifer and a water flea (*Daphnia*).

Clever Hans The horse named Clever Hans was reputed to be able to do arithmetic; when presented with a sum written on a blackboard, it gave the answer by tapping out the number with its foot. Tests proved that the animal was really responding to subtle cues, unwittingly given by the questioner, that signaled when to stop tapping. Ever since this demonstration of how unconscious cueing can affect an experiment on animal intelligence, behavioral scientists have been alert to the possibility and tried to exclude it. Some of the recent work on language learning in apes has been criticized on the grounds that a Clever Hans effect might have been involved.

Clinging young The young of primates and some other mammals (for example, koalas, sloths, bats) that cling firmly to the mother during the first days of weeks of life and are carried about by her. The grip is maintained in most cases by reflex flexion movements of hands and feet. Young bats also hold fast by their mouths to the mother's nipples ("anchorage nipples"). Primate infants often hold the mother's nipple in the mouth longer than nursing requires, although there do not appear to be anchorage organs in these cases. Young carried in this way appear to be roughly intermediate between altricial and precocial types of development: in the degree of development of sensory capacities they are comparable to precocial young, but in motility they are much less advanced, being incapable of following the mother as precocial young can. They can thus be described as passive precocials. See also CLASP REFLEX, CONTACT BEHAVIOR, TRANSPORT OF YOUNG.

Closed group See Society

Cocoon Protective envelope for eggs, larvae, or pupae. Cocoons are formed from secretions of specific glands, whose structure and location differ in the various species possessing them: spiders use the web silk glands; many insect larvae (such as ichneumon flies) use the salivary glands; and earthworms have special mucous glands in the anterior body segments. Egg cocoons are produced by the mother or hermaphrodite parent; larval cocoons are produced by the larvae themselves (for example, insect larvae at the time of pupation).

Cognition   In human psychology, a general term for mental functioning, including perception, memory, and thinking; also an individual's knowledge of local geography (see COGNITIVE MAP), awareness of self (see CONSCIOUSNESS), learning, judgment, and use of language. It does not include emotion and willing (conation, motivation).

The concept of cognition has recently found increasing use in ethology, especially in primate studies. However, its direct applicability is a matter of controversy, for whereas cognition connotes consciousness in most human contexts, only very guarded assertions can be made about the existence and nature of consciousness in animals.

Cognitive ethology   That part of ethology having to do with animal cognition.

Cognitive map   An internal representation of the layout of an area of the environment. Tolman argued that such representation is implied by the ways rats can run mazes when presented with shortcuts or detours they have not previously taken.

Colony   In biology, a congregation of individuals of the same species, "congregation" here carrying no implication about the nature of the cohesion. The individual members may be joined together as branching growths, in which case their individuality may appear to be merged in the whole, as in many of the marine animal forms described as colonies (for example, hydroid colonies, Portuguese men-of-war, sea pens). They may be held together by gelantinous secretions, as in the colonies of many algae. And colonies can be free-living, as the breeding colonies of many birds are, and the societies (sometimes described as colonies) of ants and termites, and sedentary groups of spiders. In the case of insects, however, because of the degree of mutual dependence between individuals, the question can be raised of the extent to which they have true individuality.

Color change   Alteration in the color of the body. The color of the whole body surface, may change, or of specific parts or structures. A color change may occur only one or a few times during development, may recur seasonally (see ADVERTISING DRESS), or may occur at irregular intervals. The duration of color changes thus varies from many weeks or months to a few minutes or seconds. During long-term changes the quantity of pigment in the pigment cells changes; in the quick processes only the distribution of pigment within the cells is altered. The plumage of birds changes color through molting, as a rule; similarly, the coat color of mammals may change through shedding and regrowth of hair. Molting, hair shedding, and slow color changes of the skin, such as the reddening of the ventral surface of male three-spined sticklebacks, are controlled hormonally; the quick color changes are effected more directly by the central nervous system in most cases.

The biological significance of color change varies, but its usual function is either camouflage or social communication. The remarkably rapid color changes of many fishes and cephalopods enable them to make sexual and aggressive releasers appear and disappear almost in a flash.

Luminescence causes a sort of color change by the production of light instead of its reflection. Special luminescent organs are used by many animals living in the dark or active only at night to flash social signals (fireflies) or to present lures to prey (anglerfish).

Comfort behavior   In ethological writing this term covers two very different groups of behavior patterns: those having to do with care of the body (preening, grooming, scratching, shaking, rubbing, washing, dustbathing); and those concerned with relieving tensions arising from metabolic functions (defecation, urination, stretching, yawning). The term's range of application is neither uniform nor sharply defined. For example, some ethologists include sleeping as well as selecting and preparing a place to sleep in the category of comfort movements; others do not.

Some elements of comfort behavior have evolved into components of display. A familiar example is the monstrous yawning of many primates, which serves a threat function.

Commensalism   A form of symbiosis in which one of the animals derives benefit from the arrangement but not the other, which neither gains nor loses. For example, various animals live as commensals in the nests of social insects, obtaining shelter and perhaps food from their hosts without supplying anything in return but without

49

causing any harm. The lines between commensalism and mutualism on the one side and parasitism on the other are ill-defined.

Communal courtship Communal mating; group mating; group courtship. A general term for courtship in which more than two individuals take part. (The term "social courtship" is also sometimes used in this sense, but this can be confusing, since, according to the meaning of "social" in ethology, all courtship is social.)

Two different forms of communal courtship occur in animals. In the first, males gather and display in concert to an audience of females, thus jointly enticing and stimulating the females and at the same time competing for their attention. Examples are the species of birds and mammals known for their lek displays, such as birds of paradise, ruffs, grouse, the South American cock-of-the-rock, and the Uganda kob, as well as the swarms of certain insects, (such as dancing midges). In the second form of group courtship, males and females play more or less equal parts. This occurs mainly in group-living species of the tropics, probably because the mutual stimulation facilitates reproductive onset and synchrony in the absence of an annual zeitgeber (see FRASER DARLING EFFECT). A notable example is the communal ceremony of flamingos, which consists of stretching and preening movements by dozens, even hundreds, of the birds. In addition to mutual stimulation, these gatherings may also assist animals at the same stage of reproductive readiness to meet and so form a breeding colony or part of one (Orians effect).

Communal mating See Communal Courtship

Communal song Group song; chorusing. Singing together by a number of conspecifics. Sometimes the term is also applied to the collective production of trains of calls as occurs in many frog species. Two different forms of group song are distinguished, on the basis of the contributions of the sexes. In the first type, which occurs predominantly in crickets, locusts and cicadas, frogs, flying foxes, and some birds (hummingbirds and birds of paradise), the males sing together to attract females. In the second type both males and females take part in the communal singing, which is sometimes compounded of duets. This form probably promotes cohesion of the group, synchronizes the members' physiological states or activities,

50

and possibly serves as acoustical advertisement of a group territory (see MARKING BEHAVIOR). This kind of group singing is especially highly developed in some primates, particularly gibbons and howler monkeys.

Communication Signal transmission. Influencing the behavior of another animal by means of signs or displays. To distinguish the ethological use of the term from its use in other contexts, such as telecommunication technology or sociology, animal signaling is sometimes referred to as biocommunication (compare SEMIOTICS). The study of animal communication has always been a major part of ethology.

Depending upon the sensory modality, communication may be optical (visual), acoustic (vocal), chemical, tactile, or even electrical. Optical signals comprise features of form, coloration, posture, and movement; acoustic signals consist of vocalizations and sounds produced by such means as stridulation, knocking on wood, and drumming of feathers; chemical communication employs pheromones; and electrical communication depends upon phasic production of electrical fields. Human language is the most sophisticated form of communication to have evolved. The extent to which other forms of animal communication share the design features of human language, and the extent to which other creatures might be able to acquire language are matters of considerable current interest and controversy. In this connection the distinction between verbal and nonverbal communication is pertinent. Also controversial is whether animals can use their signals for deception and whether the main function of communication is to convey information or to manipulate the recipients.

Most animal communication takes place between conspecifics. However, there are numerous cases of signaling between members of different species (interspecific communication). For example, the alarm calls of many songbird species are so similar to one another that they are understood by members of other species as well as by conspecifics—an avian lingua franca. Interspecies communication is necessary for most cases of mutualism.

Communication theory In some contexts this is identified with information theory. However, if the latter is defined as the mathematical treatment of the statistical concept of information, there is a need for a broader term that includes other aspects of signaling yet is more focused than semiotics, which covers a broad range from ethology to literary criticism. Communication theory analyzes the different forms of communication and examines the relationships between their constituents. It defines communication as the transmission of a signal via some channel from a sender to a recipient. Study of what the signal encodes about the sender is referred to as semantics; study of the physical nature of the signal is called syntactics; and study of the effect of the signal on the recipient is pragmatics. In Smith's (1965) terminology, what a signal encodes is its message, and the effect it has on a recipient is its meaning, which is a product of the signal and the context in which it arrives (see MESSAGE-MEANING ANALYSIS). A distinction is drawn also between features of a signal or sign that carry the communication (sign vehicle) and features that are merely incidental.

Companion A term used in the earlier ethological literature for a social partner in a behavioral nexus.

Comparative ethology A term signifying the direction given to ethological study by its founding fathers, Lorenz (1950) and Tinbergen (1951). In contrast to behavioristic and human psychology, comparative ethology always places species comparison in the foreground. Unlike comparative psychology, it pursues this comparison as it applies in phylogeny and evolutionary biology and also in ecology, that is, in accordance with the concepts of comparative morphology. The major contribution of this school, besides the formulation and explication of most of the original concepts of ethology, was the recognition and demonstration that behavior patterns, especially the so-called fixed action patterns, offer species markers that are just as good as morphological or physiological characteristics and that they too can be treated from the viewpoint of phyletic descent.

From the perspective of the history of science, the term was justified because it drew attention to the marked contrast between ethology and many of the psychological approaches to animal behavior. Now, however, it is little used, because a comparative perspective characterizes other branches of biology as well and so is more or less taken for granted within biology.

Comparative psychology Orginally conceived by G. Romanes (1882) as the comparative study of mind in the animal kingdom. Romanes used the "ejective" approach of trying to imagine himself in the situation of the animal, with its size, sensory capabilities, and motor capacities, and so arrive at some idea of what its subjective states must be like. This way of arguing from human to animal experience is now usually dismissed as idle anthropomorphism. Indeed, comparative psychology today would be more aptly described as comparative animal behavior, since it deals mainly with observable behavior and tends to eschew mentalistic terms like "mind," "feeling," and "intention" (but see COGNITIVE ETHOLOGY). To the extent that comparative psychology is distinguishable from ethology, it reflects a psychological rather than a biological heritage, tending to emphasize learning and the physiological basis of behavior. The term now covers a broad, ill-defined group of mainly American scientists, whose aim is to investigate differences and similarities in the behavior of different animal species. It differs from comparative ethology in its concern with aspects of behavior that are relevant to human studies and that cannot be experimentally investigated in humans for ethical and methodological reasons. Like behaviorism, though without going to its doctrinaire extremes, comparative psychology has emphasized environment-dependent features of behavior, especially of perception, habit, and learning, for which animals can be used as surrogate people. However, because the animals studied used to be chosen largely according to their suitability for studying such questions and so were virtually restricted to a few species of domesticated laboratory animals, and because the methods and questions were sometimes quite unbiological, the extent to which comparative psychology could be truly comparative in a biological sense was questioned. More recently the work of comparative psychology has broadened and turned more in the direction of ethological questioning and methodology. Hence the differences between these two branches of behavioral science have progressively narrowed, and it is now impossible to draw a sharp line between them.

Compass orientation See Orientation

Competition In biology, a situation in which two or more individuals or species lay claim to the same natural resource. The most obvious competition is that between conspecifics; in addition to survival and reproductive necessities (food, nesting places, building material, resting places), it may concern social resources (for example, mating partners or rank order status). Competition is a cause of aggression between conspecifics, but the degree of fighting is kept in check by threat behavior and intimidation; this is shown in the development of dominant-subordinant relations within groups (rank order) and the establishment of territories, which can reduce physical competition by adjusting the distribution of a population to the available living space.

Interspecific competition is also common, especially between closely related species occupying overlapping habitats. In such cases adaptations analagous to the aggression-reducing measures of intraspecific competition can arise, such as interspecific territoriality. Even in the absence of such manifest consequences, two or more species will compete wherever their niches overlap. According to the competitive exclusion principle, if two species with the same ecological makeup coexist in the same region, the one that has even a minute marginal edge in exploiting the resources will eventually drive the other out, or the two species will evolve divergent ecological adaptations and so cease to be head-on competitors.

Where competitors actively contend with one another (for example by fighting or threatening), they may be described as rivals and as engaged in rivalry.

Competitive exclusion principle "No two species can coexist at the same locality if they have identical ecological requirements" (Mayr 1963). This important consequence of interspecies competition is also known as Gause's principle.

Concept formation The highest intellectual achievement of animals; the ability, experimentally demonstrated in certain animals, to grasp simple concepts, which they abstract from and apply to a variety of instances. Thus a civet cat (Viverridae) acquired the concept "same/different": after initial training to pick out dissimilarity in a preliminary set of patterns, it immediately and consistently selected the odd patterns when presented with combinations from a test set quite unlike the training set. Similarly, a rhesus monkey, when trained with groups of three objects to choose the one that was different, applied the same principle when faced with other sets of objects differing from the initial set in both form and color. Chimpanzees and rhesus monkeys have even been taught value concepts: when offered different-colored tokens, which could be exchanged for different amounts of food, they learned to pick out the token that gave the highest return, irrespective of the arrangement of the tokens.

Concept formation of this kind implies a capacity for quite sophisticated abstraction, and so, not surprisingly, it appears to be more or less restricted to the higher mammals. Since animals are incapable of acquiring linguistic concepts, with the possible exception of some apes, dolphins, and a parrot (see LANGUAGE), their conceptual abilities are described as nonverbal or prelinguistic. The counting ability found in some species of birds also implies conceptual capacity of a sort.

Concorde fallacy   The principle that the amount of time and energy an animal has expended on a project dictates how much effort it will exert to carry the project to completion, even if the costs will clearly exceed the possible benefits. The analogy is with the Anglo-French supersonic transport aircraft, which continued to be developed even after the enterprise appeared economically and practically misconceived, because, it was argued, so much had been spent on it already. So, it has been argued, a mother is less likely than a father, to desert her family because her investment in the brood is much greater (see PARENTAL INVESTMENT). Considerations of fitness, on the other hand, argue that past investment is irrelevant to the present decision; according to this view, the animal assesses which course of action will be the most profitable in terms of inclusive fitness. Nevertheless some animals do appear to commit the Concorde fallacy. For example the vigor of the female digger wasp's efforts to defend her possession of a burrow against the efforts of other wasps to take it increases with the number of insect prey she has stocked it with, whether or not the burrow was already stocked with prey when she acquired it.

Conditioned reflex   See Conditioning

Conditioning A general term for experimental procedures that can make a behavior pattern or other reaction contingent upon certain conditions. An animal so treated and the behavior so affected are said to be conditioned to the stimulation or situation. An animal whose behavior is affected in a comparable manner in natural circumstances might also be described as conditioned.

The two standard forms of conditioning are classical, or respondent, and operant, or instrumental. Classical conditioning begins with a natural or preestablished stimulus-response connection. The unconditioned or unconditional stimulus (for example, the sight or smell of food) is presented in association with an arbitrary stimulus (the conditioned or conditional stimulus, such as a ringing bell or flashing light). After a certain number of joint presentations, the response that initially was dependent on the unconditional stimulus (the unconditioned or unconditional response, for example, salivation) is elicited even when the conditional stimulus is presented by itself; the response now is a conditioned or conditional response or reflex. Classical conditioning is sometimes called Pavlovian conditioning after the Russian scientist credited with its discovery.

Operant conditioning differs from classical conditioning in that instead of bringing a new stimulus into association with a given behavior pattern, a new pattern is brought into association with reduction of a need (allaying of hunger or thirst) or some other sought-after circumstance. The behavior, for example making a right or left turn, pecking at a colored or otherwise discriminable target, or pressing a lever, must first occur spontaneously or by accident. When reinforcement, for example food pellets for a hungry animal, repeatedly follows the behavior, the animal forms or strengthens the association between the behavior and the need state (hunger), or expectation of reinforcement, in the training situation. A further difference is that in classical conditioning the animal learns passively to associate the stimulus with the consequence, while in operant conditioning it actively performs the behavior that becomes conditioned. However, some learning theorists have questioned whether the difference in experimental procedures corresponds to a difference of learning process; they argue that even operant behavior starts under the influence of some stimulation present in the experimental situation (that is, they do not believe in truly spontaneous behavior), and even classically conditioned behavior can appear to be actively launched and infused with anticipation.

Apparatus specifically designed for work on operant conditioning includes the maze and the Skinner box (named after B. F. Skinner, one of the foremost investigators of this kind of learning). The most commonly used animals in conditioning experiments have been laboratory rats and pigeons. Operant conditioning is sometimes also called learning by result, trial-and-error learning, or shaping. Most cases of training are really cases of operant conditioning.

Both forms of conditioning practiced in the laboratory have parallels in nature. For example, free-living animals have frequently been observed to try out some new action when foraging or nest building. If some movements then prove more effective than others, and are consequently reinforced as in laboratory conditioning, they become associated with the situation.

Configurational stimulus   See Gestalt perception

Conflict   In addition to its common meaning—the clashing of opposing interests, as when two animals claim the same piece of ground—this term has a motivational connotation for ethologists. In this sense an animal is in a state of conflict when two incompatible tendencies are aroused equally, for example, attacking and fleeing at the boundary in territorial disputes. This state is expressed in conflict behavior, which includes ambivalent behavior, displacement activity, and redirection.

Consciousness   Immediate awareness of things, events, and relations. Many people consider the consciousness of animals closed to empirical investigation and therefore of no concern to ethology. Although various observations suggest that at least the more "highly evolved" species can experience some form of awareness, such as bodily feeling, statements about the nature of this awareness in comparison to that of humans are dismissed by tough-minded scientists as idle speculation. Consequently ethologists, as a rule, have avoided using concepts that suggest the presence of human forms of consciousness in animals. When ethologists use mentalistic terms (for example, individual recognition, insight, motivation) they often redefine them operationally so as explicitly to exclude connotations of consciousness. However, the recent emergence of cognitive ethology has lifted the ban on discussions of the question of animal awareness, even though upholders of the conservative position remain opposed to the new liberality with regard to contemplation of animal minds.

Consort pair In the primate literature, a temporary partnership couple within a polygamous group. In many species (baboons, Barbary apes, chimpanzees), a male and female will stay particularly close during the estrus period, sometimes removing themselves from the group a bit and engaging in much sexual activity and social grooming.

Consummatory act, consummatory stimulation See End act

Contact animals Animal species that lack individual distance. However, the division into contact and distance animals, in vertebrates for example, is hardly clear-cut. Examples of contact animals include eels, morays, catfish, chelonians, and some lizards and snakes; among birds, mousebirds (Coliidae), white-eyes (Zosteropidae), penguins, and most parrots; among mammals, most rodents, pigs, hippopotamuses, and most primates. Sometimes, as for example with tropical estrildine finches, there are differences in contact behavior even between closely related species. The reason for such differences has yet to be found. Whether or not a species is a contact animal does not seem to be related to the degree of sociality; many swarming and herd-forming animals—fish, birds, and mammals— are patently distance animals. (See DENSITY-TOLERANT ANIMALS.)

Contact behavior Huddling. In ethology contact behavior usually refers to the gathering together of animals that lack individual distance and hence are comfortable when touching one another. Some contact animals seek the greatest possible area of mutual surface contact. Probably the primary function is protection against cold, but in many species contact behavior serves in addition or instead as a component of pair or group formation. Contact behavior is especially strongly developed in the primate infants referred to as clinging young. If these animals are deprived of close bodily contact with the mother in early life, as can happen if they are hand-reared, they will have serious deficiencies in behavioral development (see DEPRIVATION SYNDROME).

In a broader sense, contact behavior refers to the exchange of signals (for example, vocalizations) in socially living animals. Many birds use contact calls when moving as a flock through vegetation that interrupts visual contact, thus ensuring that the flock remains intact.

Context The situation in which a signal occurs, which can influence the effect of the signal on a recipient. For example, the song of a male songbird will attract an unmated female in a courtship context but will repel a rival male in a context of territorial defense. The context includes everything besides the signal influencing the effect on a particular occasion. It can include, as in the example given, the sex of the recipient and the other signals given in sequence or in concert with the signal in question (see COMMUNICATION THEORY, SYNTAX).

Contingency Spatial or temporal association. The joint (simultaneous or sequential) occurrence of two behavior patterns, or a stimulus and a behavior pattern, or a behavior pattern and reinforcement, such as can give rise to persistent linkage (see LEARNING). The contingency of reinforcement plays an important part in operant conditioning; some would say it has an essential, if not defining, role. Also under natural conditions temporal relations between stimulus and response can be of considerable significance, especially in social interactions. Among some birds, the unhatched young learn prenatally to distinguish the calls of their own parents from those of neighbors, apparently because the chicks' parents immediately answer the chicks' calls, and no such contingency occurs for the vocalizations of the neighboring birds.

Control theory That part of cybernetics dealing with automatic regulation of machines, including organisms. It is highly general in application, hence abstract and mathematical. Its influence in ethology is most prominent in theorizing and research on feedback mechanisms in animal behavior.

Convergence Convergent evolution. In biology one speaks of convergence when the same or similar ecological conditions lead to acquistion of comparable characteristics in different species independently. As they adapt to these conditions, such species grow more and more alike, at least superficially; that is, they evolve analagous features. In the realm of behavior, eggrolling and distraction displays have evolved convergently in numerous unrelated ground-nesting bird species. Also the mouth breeding of various fishes (see MOUTH BREEDER), the sucking manner of drinking in pigeons and some estrildine finches, and the ability to hover on the wing shown by many insect and bird species are results of convergence. The numerous morphological, physiological, and ethological correspondences between the Australian marsupials and the placental mammals offer an especially impressive set of examples (see ADAPTIVE RADIATION). The study of convergent adaptation is a prime concern of behavioral ecology.

Similar to convergent evolution is parallel evolution, in which related species are independently subject to similar selection pressures and consequently evolve along similar lines. Thus different species of African antelopes possess many antipredator adaptations, both morphological and behavioral, in common, as a consequence of evolving in similar habitats in which they have been prey to the same carnivores since they split from their common ancestors.

Coolidge effect   The effect of novelty on the sexual activity of male animals. When a male has copulated with a female to the point of showing no further sexual interest in her, replacing her with a new female can stimulate him to resume copulation, at least until his sexual activity again wanes. The sociobiological explanation is that this effect is a means of maximizing the male's individual fecundity: once copulation with a particular female has led to fertilization, and hence engendered as many offspring as the occasion allows, it would be a waste of time from a fitness point of view to continue; for further sexual activity to be effective in this sense it has to be with another, unfertilized, female. However, there is evidence that the proximate cause of the sexual rearousal of a male may simply be novelty or change in the circumstances; reoffering the same female in a different setting can have an effect comparable to offering a new female in the old setting. Numerous cases of the Coolidge effect have been found in mammals and fish.

Copulation   Mating; sexual coupling of male and female. In mammals and some representatives of other vertebrate classes the male copulation organ (penis) is inserted into the female aperture (vagina) during copulation. A variety of analogous forms of copulation are found among the invertebrate phyla. In most birds, which lack a penis, the male presses his cloaca against that of the female and, at the climax of copulation, ejaculates sperm into her reproductive tract, which leads to internal syngamy and hence fertilization. Many lower aquatic animals do not copulate; they merely discharge their gametes into the water in situations conducive to external fertilization.

In mammals copulation consists of the male mounting the female, usually from behind; intromission, which is the insertion of the penis into the vagina; pelvic thrusting and ejaculation, the forceful discharge of semen into the vagina. Species differ in the numbers of mounts, intromissions, and ejaculations making up a typical copulation bout and also in the temporal characteristics of the behavior. Carnivores usually form a tie after ejaculation, which can hold the two animals together for an hour or more. This, like the sperm plugs formed by numerous other kinds of mammals, no doubt helps to ensure fertilization by preventing sperm from spilling out.

Counter singing   See Duet singing

Counting ability   As determined experimentally by instrumental conditioning, the capacity of many birds and mammals to discriminate numerically, for example, learning to open a specific number of food containers or to cease pecking at grains in a food dispenser after taking a certain number. The upper limit in numbers that animals can handle when tested in this way has proved to be six for magpies, ravens, and squirrels; six to seven for budgerigars and jackdaws; and eight for pigeons. In other experiments, the animals are offered patterns with different numbers of marks, which they discriminate numerically.

Animals, except for some apes possibly, are incapable of developing a linguistic concept of number, but they can be described as having an inarticulate or preverbal capacity to count (see CONCEPT FORMATION, LANGUAGE). Training in counting is one of the methods by which ethologists have investigated the higher mental capacities of animals.

Courtship Overtures to mating. For an ethologist courtship in a narrow sense covers all patterns of behavior that initiate mating or aim to do so. However, because such behavior does not always lead to copulation but rather furthers pair formation or consolidation of pair bonds, and since its demarcation from other pair-formation behavior is often vague, the concept of courtship is frequently more broadly conceived, encompassing all behavior patterns of precopulation, pair formation, and pair bonding. Consequently scientists may use this term in very different ways, and it is advisable for a scientist to state at the beginning of a publication on courtship behavior precisely what the term is intended to cover.

Courtship behavior, in the wider sense, serves the following functions: (1) meeting of the sexes (which sometimes live apart even during the breeding season); (2) overcoming of attacking and fleeing tendencies and individual distance; (3) temporal fine tuning of the partners (see SYNCHRONIZATION) so that both are ready for copulation, and for any subsequent joint parental behavior, at the same time; and (4) prevention of interspecies hybridization (see ISOLATING MECHANISM, REPRODUCTIVE ISOLATION). For this last, courtship behavior must ensure, usually by means of releasers, that only the right (conspecific) sex partner will be attracted. The more closely related the species in a region are, the more elaborate and hence the less easily mistaken the courtship displays and associated behavior tend to be (as is the case for birds of paradise and for ducks). Not all of the behavior patterns of courtship contribute equally to these functions; different action patterns may be responsible for different functions (for example, attraction or synchronization).

In reference to mammals, it is more customary to speak simply of mating behavior, or in the exclusively mammalian terms of rutting (males) and heat (females), which, however, carry the additional connotation of physiological readiness for mating. Some terms, mainly from the language of hunters and gamekeepers apply to particular species. Precopulatory behavior is sometimes described as foreplay, a term that is perhaps ill-advised since, in contrast to true play behavior, the acts are generally carried out in earnest. Compare POSTCOPULATORY BEHAVIOR.

Courtship chain   See Action chain

Courtship feeding Feeding the mate. The actual or sham giving of food to the partner as part of courtship or pair bonding. It is especially widespread among birds. Originally the male probably fed the female to provide her with all of, or a supplement to, her food supply during egg production and brooding. No doubt it still serves this function for many species. In many cases, however, courtship feeding has undergone an extension or complete change of function, serving additionally or exclusively as a component of courtship or for strengthening the pair bond. In such cases actual food may no longer be transferred, the "feeding" being merely symbolic, as in billing. If food is supplied, courtship feeding may provide a female with a means of judging the male as a potential father and caretaker, as has been shown in common terns. In addition to birds, insects and spiders furnish many examples of courtship feeding.

Courtship flight A term frequently used for the song flight of many birds. Especially among birds that live in open landscapes where there are not enough bushes or trees to provide singing posts, the singing is frequently performed on the wing. The most familiar example in Europe is the skylark. In North America it is exemplified by the nighthawk.

The term "courtship flight" is misleading in that most song flights are not part of the initiation of mating, but serve first and foremost for a male to advertise his territory (see TERRITORIAL SONG), both to warn off other males and to attract females. More accurate terms would be advertisement flight or territorial flight, or, quite neutrally, display flight or song flight. Courtship flight could then be reserved for the few bird species in which the call or song flights serve solely to solicit females as a prelude to mating. Examples are the display flights of various heron species and the joint pairing flights by males and females in many hummingbird species.

In some cases vocalizations other than song occur in flights that serve one of the functions of song flight, such as the calls of many plovers and waders, and even nonvocal sounds, such as the clapping together of the wings of pigeons and the "drumming" produced by the flow of air over the tail feathers of diving snipe. In social insects the swarming flights of males and females that precede pairing are referred to as nuptial flights. Insects also offer examples of individual courtship flight, such as the flashing flights of male fireflies, which serve to attract mates.

Courtship song   In the ornithological literature this term is used for songs that serve exclusively to attract or stimulate females, thereby contributing to courtship, unlike most birdsongs, which have the double function of territorial marking and female solicitation. Such songs occur in, for example, estrildine finches, bullfinches, and the grassquits of the Caribbean. The female-attracting songs of insects are usually referred to as luring songs or soliciting songs.

Creche   In the ornithological literature, a group consisting of the young of several families tended by one or more adults (usually fewer than one parent from each family). In a narrower sense, "creche" is the term for gatherings of precocial birds in which the young can forage for themselves and for which the adults only have to provide guidance and defense against predators. The tending adults are either parents of the young birds or nonbreeding adults or those that have lost their broods. Here parental care may be shown toward young that are not related to the adults. The best-known example of this form is found in eider ducks. The benefit to the mother probably is that she can leave her young long enough to forage and replenish the weight lost during incubation. Creche formation of this kind may be a case of reciprocal altruism.

In a wider sense the term is applied to assemblages of altricial young that have left the nest or brood place yet are still fed by adults. Only a few adults are present (and sometimes none for a spell), but in contrast to the previously described form, the parents regularly return and feed their own young. In some birds, such as penguins, guards attend the group of young but do not feed them. The guards may be nonbreeding adults. Creche formation of this kind has been described for penguins, flamingos, pelicans, terns, and cockatoos, and jays.

Similar to creches among birds are the gatherings of young animals of many ungulate species with a lying-out period (see entry). These are sometimes described as kindergartens.

Critical period   See Sensitive period

Cronism Coined from Cronus, the Greek god who devoured his offspring, this term refers to a parent's eating its young. If the young are killed but not eaten, the killing is spoken of as infanticide, but this term frequently includes cronism as well. Cronism is a special form of cannibalism. White storks have quite often been reported as devouring their young, apparently because of the inadequate development of parental behavior in pairs breeding for the first time. Under laboratory conditions cronism can occur as a consequence of social stress. It is much more frequently observed as a behavioral disturbance in domestic animals (such as the sow who eats her farrow) than in wild animals.

Cross-cultural comparison See Human ethology

Cross-fostering See Foster rearing

Crypsis See Camouflage

Cultural inheritance See Tradition

Culture In ethology, a term that is used only sparingly. Phenomena comparable in some respects to human culture are more or less confined to the primates. Several cases are on record of individual primates discovering new ways of doing something, which are then copied by others. Such imitated novelties may be described as social modifications. The best-known example is the potato washing of Japanese red-faced macaques. When a behavioral variant of this sort spreads within a group, when it is passed on from generation to generation through tradition, and when this group persistently retains it but only as an acquired behavioral trait, then we have culture in the ethological sense. In the words of Kummer (1971): "Cultures are behavioral variants induced by social modification, creating individuals who will in turn modify the behavior of others in the same way."

In biology "culture" has another sense, which ethologists also sometimes use: the medium in which microorganisms or cell types are cultivated. For example, there are tissue cultures, the agar gels useful for growing bacteria, and the infusions that can become popu-

65

lated by infusoria, such as protozoa. In this connection the word can be used as a verb.

Curiosity   See Exploratory behavior

Cut-off   A means by which animals in situations of social conflict reduce stimulation that tends to make them distressed or makes them behave in ways contrary to what they are aiming at. For example, looking away or averting the eyes from another individual that is arousing fear or hostility is a cut-off that, enables the animal to contain its tendency to flee or attack when it is trying to maintain its ground against a rival or establish a relationship with a prospective mate. The most obvious examples are the cutting off of visual stimuli, but a person's covering his ears with his hands so as not to hear something embarrassing might count as a case in another modality. The term was introduced by Chance (1962), who applied it to such patterns as the facing away of gulls.

Cybernetics   The science of control and communication in animals and machines. It comprises two divisions: control theory and information theory. Cybernetics deals with the general principles and theory of automata and information transmission. Originally developed in engineering, it has found wide application in biology, including ethology. For example, the concept of feedback, although anticipated in earlier work, was formulated by cybernetics in a way that led to many subsequent applications in biology. Ethologists owe the mathematical definition and measurement of information to cybernetics.

DIMORPHISM

D

Damming up An older ethological conception of how readiness to perform a particular behavior pattern may increase with the length of time since the last performance. For example, an animal was said to be sexually dammed up if some considerable length of time had passed without opportunity to engage in sexual behavior. Outwardly the condition can be detected as threshold change. The metaphor was linked to the hydraulic models of motivation developed in classical ethology and, like those models, is now rarely used.

Dance language A form of communication by honeybees, which is comparable to language in that it depends partly on the use of signs to convey information. A worker bee returning to the hive after foraging in flowers can inform its companions about the direction and distance to the food source through a series of movements referred to as a dance. If the food source is in the neighborhood of the hive, the bee performs a round dance, running in circles on the honeycomb in a rapid and conspicuous manner, which attracts the attention of other workers and induces them to follow her movements. The operative information is in the flower scent adhering to the signaler, which her hive mates perceive as they contact her with their antennae and can track to its source when they fly out of the hive. If the food source is farther away, however, the directional and distance information is transmitted by movements of the dance itself, which has a much more definite form and is referred to as the waggle dance: the bee describes a flat figure of eight, and during the straight run between the circlings back, it waggles its abdomen from side to side. In this the bee accomplishes a remarkable feat of translation. The direction of the food site relative to the direction of the sun's position, which cannot be directly represented on the vertical surface of the comb, is indicated by drawing on the direction of gravity: if the waggle run is directed vertically upward, the food source lies in the same direction as the sun; if it deviates by a specific angle to the right or left of vertical, the direction to the food is at a corresponding angle to the sun's position. In addition, the distance to the food is encoded in the speed of the waggle run: the farther away the food, the slower and more emphatic is the waggle run. However, bees following the directions of a waggle dance can also use visual and olfactory cues as aids to finding the food source, especially for more distant sites.

The dance language is used also when bees have to find a new site for a hive. Returning scout workers inform their colony companions of suitable locations they have discovered in their explorations.

Deafferentation   Surgical severing of the afferent nerves to eliminate external stimulation and allow investigation of the spontaneous component of behavior (see AFFERENCE, SPONTANEOUS BEHAVIOR). However, because deafferentation can lead to degeneration of cells in the region of the brain to which the severed nerves project, results obtained with this procedure must be interpreted carefully.

Deafferentation has been especially useful in work on bioacoustics: by surgically deafening young birds, scientists have been able to determine which components of a bird's vocal repertoire are independent of experience, which have to be learned, and at what ages such learning can occur (see SENSITIVE PERIOD).

Death feigning   See Thanatosis

Deception Misinforming or disinforming. The use of behavior or appearance, signs or signals, to create a false impression. In ethology and sociobiology the term applies to an animal's taking another organism to be something other than what it really is, whether or not there is any intention to deceive. Thus when a wasp attempts copulation with the orchid *Ophrys insectifera* because the flower looks like a female of his species, it can be said that the wasp was deceived by a design of nature. Other instances of mimicry (with the exception of Mullerian mimicry) can also be regarded as deception in this sense. Behavioral examples of deception include the "femme fatales" of the firefly *Photuris versicolor*, which lure males of other species into their predatory clutches by flashing in answer to the males' flashes; the distraction displays with which ground-nesting birds such as ducks and plover attempt to lead a predator away from their clutch or brood; and, if the Beau Geste hypothesis applies (see SONG REPERTOIRE), a bird's use of a large song repertoire to create the impression that several birds are established in the area it is defending and so deter territory-seeking rivals from attempting to settle.

The question of intentionality does arise for some cases of deceptive behavior, especially in birds and mammals. Some performances of distraction displays are so cunningly controlled as strongly to tempt an observer to conclude that the birds must know what they are doing; this possibility is being put to experimental test. In the case of many primates the impression of intent is even more irresistible.

Decision making Choosing between behavioral options. Animals are often faced with a choice between alternative courses of action, for example, staying to protect a brood or fleeing to escape destruction when threatened by a predator; holding ground or conceding defeat in a struggle over territory; continuing to provide care for present young or weaning them and starting another family. Ethologists have pondered what dictates the animal's choice in such cases. One approach is to consider what best serves the animal in terms of fitness. For example, on the question of a parent risking its life for its brood, the answer depends on how much the parent has to lose: if the animal is in the first of many breeding seasons, theory predicts that it will put its own survival ahead of that of its young, for they represent fewer progeny than the parent would produce by realizing all of its breeding potential. But if the animal is at the end of its breeding life, or if it breeds only once, it should do all it can to save the existing offspring. Similar considerations of parental investment bear on questions of mate fidelity and the timing of weaning. Other ways to assess the costs and benefits in terms of fitness or "energy budgeting" include optimization theory and the games theory of evolutionary stable strategies.

Aside from the "economics" of decision-making there is the question of proximate means by which the animal chooses between alternatives. Since an animal can usually do only one thing at a time, conflict or competition may occur over which of two or more motivated activities will take command (see CONFLICT). In such situations one of the rival activities may take precedence over the other as long as its causal factors are above a certain threshold (see TIME SHARING). In other situations, especially where conditions are relatively constant or predictable, selection among activities may follow a preset rhythm (see PERIODICITY). When animals have several sources of food available, their choices may be governed by a search image and hence be affected by experience of relative abundance. However, the decision strategy to which such experience is applied varies for different kinds of animals. For example, when fish are presented with a choice between two stimuli, one of which rewards response in 60 percent of the trials and the other in the remaining 40 percent, the fish distribute their responses according to the 60–40 reward ratio. Rats presented with a like situation direct all their responses to the stimulus rewarded most often, the strategy that brings the higher return.

As with numerous other terms drawn from human contexts, "decision" in ethological contexts usually does not have the mentalistic implication it has in everyday language and is therefore of metaphorical extraction.

Decremental conduction   Loss of nerve impulses from a volley due to expenditure of impulses in synaptic transmission. In a coelenterate nerve net, for example, volleys diminish with distance from the site of stimulation because of the succession of synapses (see SUMMATION).

Defensive behavior   In the ethological literature, measures taken by animals to avoid falling prey to predators. However, animals may also act defensively in agonistic behavior with conspecifics. Defense against predators is categorized as either primary or secondary defense. Primary defenses are continuously maintained, whether danger threatens or not. They include living in burrows, constructed tubes, or shells into which the animal can withdraw; camouflage and mimesis; warning of noxious or dangerous qualities (aposematic advertisement); and Batesian mimicry. The latter two are exemplified by the aposematic monarch butterfly, whose bright colors signify, to experienced birds, something to avoid, and the viceroy butterfly, which looks very like the monarch and is likewise left alone by birds, even though it is palatable. Secondary defenses are actions an animal takes when faced with a predator. They include escape behavior, including death feigning (thanatosis), protean behavior (see entry), bluffing (deimatic behavior) and deflection of attack, and various forms of defensive attack. An example of deimatic behavior is the eye-spot display of many moths and butterflies: when in danger of falling prey to a bird, the insect suddenly exposes the eyelike patterns on its hind wings, which can frighten the bird away, presumably because the pattern resembles an owl. Deflection of attack involves diverting the predator's attention to a part of the body that can be dispensed with so as to increase the chances of successful escape (for example, the false head on the rear edge of the wings of hairstreak butterflies). Defensive attack is exemplified by the mobbing of songbirds to drive owls away.

Deimatic behavior   See Defensive behavior

Delousing   See Social grooming

Deme   See population

Demography   Study of the structure and dynamics of populations, including age and sex distribution, patterns of immigration, emigration, and migration; birth and death rates; and kinship structure. Demography is a central part of population biology and hence of sociobiology.

Density dependence Effects correlated with population density. For example, when food supply is a limiting factor, the likelihood of starvation increases as population increases; also, the magnitude of social and physiological stress is a function of density. Competition, parasitism, predation, and disease are all considered density-dependent factors. They contrast with density-independent factors causing death, such as weather, the severity of which does not vary with the size or concentration of the population.

Density-intolerant animals See Density-tolerant animals

Density-tolerant animals Species with relatively high reproductive rates, which are maintained even when population densities approach overcrowding (see *r*-SELECTION). Dispersal occurs only when crowding exceeds a certain limit. Between the periodic dispersals there may be intervals of years during which numbers build up. Since crowded conditions imply considerable tolerance of bodily proximity, density-tolerant animals are generally also contact animals.

In contrast, density-intolerant animals breed at slow rates (see K-SELECTION) and tend to be territorial and solitary, or to live as isolated pairs or families. They do not show the large oscillations in population density of density-tolerant animals. Dispersal occurs regularly, usually at the end of each breeding season. Density-intolerant animals are also distance animals. However, the two extreme population and dispersal patterns grade into one another via intermediate forms.

Dependent rank See Basic rank

Dependent Variable In an experiment, the effect that is measured when the independent variable is varied. For example, hours of water deprivation for an experimental animal might be the independent variable; rate of bar pressing for water the dependent variable. Often one or more additional factors, referred to as intervening variables, have to be introduced to express the relationship between the independent and dependent variables as an equation.

Deprivation experiment See Isolation experiment

Deprivation Syndrome A general term for all the deficits and distortions in behavioral development that can result from social isolation in early life, including apathy, motor restlessness, rigid and endlessly repeated movements (stereotypy), and incapacity for normal social interaction. In some species the deprivation syndrome is especially severe for the young raised in the absence of their mother. During certain phases of development separation from the mother may have particularly strong and persistent deleterious effects. The timing of such periods of maximum sensitivity, which compare to the sensitive periods of imprinting, vary from species to species.

The deprivation syndrome and its causes have been most intensively studied in primates and other socially living animals. The concept also has application in human psychology. If the mother is removed after the bond with the infant has been formed, the effects may be referred to as separation syndrome or separation trauma.

Derived activities See Ritualization

Detour behavior See Innovation

Dialect In ethological and bioacoustical literature, a variant of a species' sound production characteristic restricted to the population of a particular region and differing from those of other regions. A dialect that has a large range or is widely distributed is described as a regional dialect; if its occurrence is more restricted it is considered a local dialect. Sometimes two or more dialects are distributed in a mosaic fashion.

Dialects are best known from the song of songbirds, but they also occur in the songs and calls of other kinds of birds. There are even dialects in nonvocal sound production, as in the drumming of woodpeckers and the wing clapping of African clapper larks (*Mirafra*), which may be referred to as instrumental dialects, as opposed to vocal dialects. Dialects also have been reported for other animal groups, such as crickets, frogs, and some mammals: whales, sea elephants, some rodents, squirrel monkeys, and langurs.

On the biological significance of dialects, ethologists' views differ widely and are, for the most part, speculative. In various bird species, such as chaffinches and ortolan buntings, dialects appear to be connected with differences in habitat. Another likely possibility is that dialects promote reproductive isolation between populations adapted to differing ecological conditions by biasing individuals toward others of their own "adaptation type" during pair formation. This conjecture is especially plausible for species with a mosaic distribution of dialects, such as the white-crowned sparrow of North America, in which races with differing adaptive characteristics and different dialects winter in the same region. Finally, it is possible that some dialects have no adaptive significance, being simply a consequence of chance variation and geographical separation. Such divergence is to be expected where learning plays a prominent part in the development of acoustic signaling.

The song dialects of birds are largely learned and are transmitted through tradition. In contrast, the dialects of crickets are apparently innate. For other groups the developmental basis of the dialects is not known.

Recently the concept of dialect has been extended to nonacoustical characteristics that show regional variation. Thus the term "chemical dialect" is used for the regional variations in the sex attractants of many butterflies or in the glandular secretions used for territorial marking of many mammals. There are also "play dialects," as exemplified in American big-horned sheep, whose young show differences in play behavior from one place to another.

Dimorphism The existence of two different morphological types or behavioral repertoires within a species or population. Two forms of dimorphism are recognized: sexual, when males and females differ in appearance or behavior, and age, when the differences are related to age. The existence of more than two phenotypes within a species is called polymorphism. See also LARVA.

Disinhibition hypothesis See Displacement activity

Displacement activity An unexpected, seemingly irrelevant movement that occurs out of the behavioral context to which it is assumed to belong functionally (that is, for which it was presumably evolved). Displacement activity occurs when an ongoing activity is thwarted in some way or when two incompatible tendencies are activated at the same time and hence are in conflict (see CONFLICT). In such situations the displacement behavior appears to be pointless because it serves neither its own function nor that of the stalled activity.

Examples include displacement pecking by domestic fowl when fighting; displacement fanning by courting sticklebacks; and displacement collecting of nest material by gulls when denied access to the nest for incubation.

The term derives from the hypothesis that when two mutually antagonistic tendencies, such as attacking and fleeing, block one another, the motivational energy jumps over into the pathways of another functional system. This has been described as the surplus hypothesis. The term "displacement" was thus both descriptive and explanatory: it denoted behavior displaced from its normal context and connoted a motivational mechanism of displaced energy. This conception implied a distinction between autochthonous causation, when a movement is caused by its "own" factors, and allochthonous causation, when the movement is caused by outside factors (see ALLOCHTHONOUS DRIVE). However, the surplus hypothesis raised problems (see MOTIVATION), which led to alternative conjectures such as the disinhibition hypothesis. According to this, the displacement behavior is caused by the same factors that give rise to the behavior at other times; the conflict or thwarting situation simply interrupts the suppresing effects of the blocked tendencies on the causal system governing the irrelevant action. In other words, the conflict or thwarting does not make the animal perform an activity categorized as displacement; it allows the animal to do so, and the distinction between autochthonous and allochthonous causation no longer applies. Other explanations have been proposed, and it is quite likely that behavior perceived as displacement has a variety of causes.

Displacement activities are believed to be one of the main sources of display behavior, the change of function brought about by ritualization. Ritualized displacement activities are especially common in courtship and threat behavior.

In experimental psychology the term "adjunctive behavior" has been used for the kinds of behavior that ethologists call displacement. In Freudian psychoanalysis the notion of sublimation has some similarity to the ethological notion of displacement. The Freudian use of the term "displacement" has more in common with what ethologists call redirection.

Display Behavior having a communication function; expressive behavior; signaling behavior; advertisement behavior. Display behavior regulates social life by influencing behavioral disposition or eliciting some response in other individuals. Display thus includes all behavior patterns that are specially differentiated to serve intra-species (and sometimes interspecies) communication, such as court-ship, threat and appeasement postures, and begging by young. Some displays are designed to deceive rather than inform. Examples include distraction displays, such as the broken-wing ploy used by many species of ground-nesting birds to lure predators away from the clutch or brood, and deimatic (defensive) behavior, such as the eye-spot displays used by many moths to startle attacking predators.

The differentiation of display behavior has taken a variety of evo-lutionary courses. In some cases it originated primarily as a social signal, as was probably the case for the showing of the swollen abdomen by courting female sticklebacks. Or it may derive from functionally different behavior through ritualization. Thus intention movements, displacement activities, ambivalent behavior, redirected movements, and surface signs of autonomic effects have frequently been starting points for the evolution of derived activities that serve as signals.

Display flight   See Courtship flight

Display territory   An area used by one or a number of animals of one sex (usually males) solely for courtship and mating. For the communal case see LEK.

Distance animals   Hediger (1941) drew attention to the fact that animal species differ with respect to their tolerance of the proximity of conspecifics. Distance animals are those that maintain a certain distance from others of the same species, that is, they avoid bodily contact as much as they can (except during copulation or suckling of young). They include salmon, trout, pike, flamingos, gulls, swal-lows, most raptors, and most ungulates. See also CONTACT ANIMALS; DENSITY-TOLERANT ANIMALS.

Distraction display   In many bird species, especially those that are ground-nesters, incubating and brooding parents react to the ap-proach of a predator by behaving in a way that is likely to lure the predator away from the eggs or chicks. The behavior consists of a complex sequence of acts, often emphasized by the showing off of contrasting patches of feathers, the whole creating a very distracting impression. A common form of distraction display is "playing the broken wing," in which the bird simulates injury, limping and dragging a wing as though it were crippled and therefore easy to catch. Another form, shown by some plover species, is the "rodent run," in which the bird mimics the movement of a mouse or rat. The bird thus diverts the predator's attention and draws the prey-catching behavior toward itself, but stays just out of reach of the predator, which it leads farther and farther from the nest or brood. When the danger is past, the parent returns to the nest. The components of distraction display appear to stem from attack and escape behavior in most cases and to be considerably ritualized. Situations that elicit distraction displays arouse conflicting tendencies, and their performance is frequently interspersed with displacement activity. Along with the often stereotyped character of the sequences, this is consistent with the theory that the displays are blind or mindless responses to circumstance, rather than deliberate strategies to deceive the predator. However, cognitive ethologists have considered the possibility that distraction displays may involve intentionality. Some experiments have attempted to determine if a bird can adjust its bluffing behavior according to the predator's reaction—its appearance of being taken in by the deception.

Diurnal rhythm   See Circadian rhythm

Division of labor   A term widely used in biology for the parceling out of functions among parts of the body, stages in the life history, or individuals within a colony or society, that are, at least to some degree, specialized for fulfilling those assignments. In the study of behavior, division of labor most often refers to the distribution of activities within a social organization. This is especially well represented in the social insects, in which different activities are carried out by individuals of different body type (see CASTE), sex, or age. In the case of age differentiation, an animal may have several roles in the course of its life, as is true for honeybees. In vertebrates division of labor is most evident in the form of leadership and caretaker roles among group-living mammals. Many vertebrates that take care of their young practice some division of labor between the sexes; for example, females may provide all of the parental care, with males defending the territory.

The functional specialization and localization of function, for example as they concern the neurophysiologist, merge with the notion of division of labor.

Domestication   The changes that take place in an animal's behavior and disposition as a consequence of being kept, raised, or bred as a farm animal or house pet or in some other captive condition governed by human management. Domestication includes the substitution of humanly controlled selective breeding for natural selection, and this can result in breeds that would be unable to survive in the wild. As a rule, species subject to such changes in the direction of selection differ from their free-living ancestral forms in having characteristics connoted by the word "domestic" (see DOMESTICATION TRAITS). The term is used for both the effecting of change and the resulting condition.

Domestication traits   Hereditary changes in behavior, body form, and physiology resulting from domestication; the ways in which domesticated animals differ from their wild ancestral relatives. Typical domestication traits include changes in body size, shape, and color (for example, albinism, piebalding); a general decline in brain weight; and changes in hair or feathers (Angora hair, curling feathers). Common effects on behavior include changes, either increases (hypertrophy) or decreases (hypotrophy), in the prominence of specific kinds of behavior. Most domestic animals are hypertrophic as far as sexual activity is concerned (see HYPERSEXUALITY); in contrast aggressiveness and parental care tend to be more weakly developed than in the wild form. Besides such differences in frequency or intensity, behavior may appear changed in its stimulus tuning (see RELEASING MECHANISM): selectivity is commonly lost in the course of domestication.

Observation of some very recently domesticated species, such as the budgerigar, which has been bred as a cage bird for only a few decades, has shown that signs of domestication may appear after only a few generations. The kind and degree of domestication effect, and hence the magnitude of differences between domestic and wild forms, depend upon the direction and degree of divergence of domestic selective breeding in relation to natural selection.

Dominance   This term has specific meanings in three branches of biology. In genetics dominance exists between two alleles when the dominant one commands phenotypic expression in the heterozygous condition; the other allele, termed recessive, is phenotypically expressed only in the homozygous condition.

In ecology dominance refers to the relative abundance of a species in comparison with other species in a biological community (see BIOTOPE).

In ethology dominance refers to superior position in a rank order or social hierarchy. An individual to which another consistently gives way is said to be dominant in the relationship, the other being subordinate or subdominant. In a linear hierarchy each individual is dominant over those below it, and subordinate to those above it, with the exceptions of the alpha animal, who is subordinate to none, and the omega animal, who is dominant over none. However, social hierarchies may be more complicated, as when animal A dominates B, B dominates C, and C dominates A. Thus dominance is relative to individual relationship. Further complications arise if two or more individuals form an alliance and so collectively are dominant over those that could dominate each of them singly. Such arrangements are common in primate groups, where social organization may also include subdivision into classes according to sex and age status, dominated by a central hierarchy of the older males. Dominance status is anything but static in the course of group life, as lower-ranking animals continually test the ability of those above them to hold their place, and as the struggle of higher-ranking animals to maintain position becomes more difficult with advancing age. Dominance ranking may vary with context, so who gives way to whom may not be the same in contests over spatial claims, food, and sexual access. There are thus problems of definition and assessment of dominance in social relationships as studied by ethologists and sociobiologists.

Dominance hierarchy   See Rank order

Drifting   See Locomotion

Drive Motivational state; urge to behave in some particular manner. The term has been used in various ways in psychology and ethology, often in one of the senses of "instinct," with which it shares a reputation for causing mischief because of its multiple meanings. In psychology drive has been invoked to account for learning; in this sense it means the mobilization of action as a consequence of bodily need, such as food or water deficit. The reduction of the deficit reinforces and strengthens the associated stimulus-response connection. In an experimental context the operational definition of drive might be hours of food deprivation, rate of bar-pressing, or some quantitive factor relating the independent and dependent variables in an equation.

In ethology the term "drive" has much the same history as motivational energy; initially it was thought of as an endogenous force propelling behavior, then was operationally construed as measurable in terms of overt behavior, and now is more or less avoided except in contexts that do not pretend to theoretical precision. Sometimes a distinction is drawn between "drive" and "urge": drive being the latent readiness to act in a particular manner, and urge being manifest motivation as expressed in behavior. "Mood" is sometimes used in a sense that overlaps with one of the senses of "drive."

Drive reduction According to some theories of learning, called drive reduction theories, reinforcement consists in the removal of a need state, and hence reduction of the drive associated with it, such as hunger or thirst.

Duet singing In the ethological and bioacoustic literature, sound production by two individuals that is "temporally and thematically coordinated in a sequential manner" (Wickler 1980). Duet singing, or duetting, occurs mainly between partners in long-lasting monogamous pairs of, as a rule, monomorphic species. It is known in mammals (for example, squirrels and gibbons), in birds (mainly passerines, but occasionally also in nonpasserines such as the little grebe and the spotted crake), and field crickets. The contributions of the two partners may be different in form and timing or more or less similar. In the simplest cases the two partners sing in alternation (see ANTIPHONAL SINGING). In more complex instances the two birds sing together, at least for part of the performance. The interval between one bird's utterance and the other's answer can be extremely short—no more than a fraction of a second. Like the latency of answering, the forms of the duet elements are often rigidly fixed through ritualization and thus show a high degree of stereotypy. Many duets consist of very complex sequences of a number of different elements.

The biological significance of duet singing is not entirely clear and probably differs from species to species. In the literature no fewer than nine possible functions are discussed. Undoubtedly it serves to keep the partners of a pair together and synchronized with one another, especially for species living in dense vegetation and for tropical species where environmental factors do not regulate circannual periodicity. At the same time duetting may serve both for territorial marking and—where it is pair-specific—for individual recognition. Not quite so highly elaborated as the pair duets is the counter singing or "song dueling" that sometimes occurs between birds and their territorial neighbors, yet this too can be comparably coordinated in timing and theme. The same is true of the alternating songs of rival field crickets.

Dummy   See Stimulus model

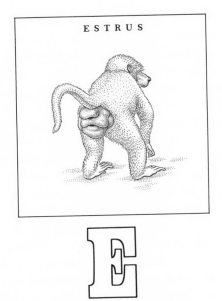

ESTRUS

Echolocation   Echo-orientation, sonar, Objects—obstacles, prey, or conspecifics—are preceived through reception of the pattern of echoes of emitted sounds reflected from the objects. Bats have the most extensively developed echolocation capacities. It occurs also in whales, especially dolphins, and in South American oilbirds (*Steatornis*), Asiatic cave swiftlets (*Collocalia*), and possibly shrews. The directional sounds of most echolocating bats lie in the ultrasonic range, above the frequency limit of human hearing.

Eclipse plumage   See Advertising dress

Ecoethology   See Behavioral ecology

Ecological community   See Biotope

Ecological niche   A term used in ecology for the position occupied by a species in the economy of its ecological system, that is, the specific role it plays within the ecological community; the living space it occupies in relation to those occupied by other members of the community. "Niche" does not have a spatial connotation but refers to a system of relationships between species and environmental conditions. Biologists differ quite widely in their definitions and uses of the term. For example, instead of emphasizing relationship to habitat and other species in the community, some authors define niche in terms of the range of environmental variables—climatic, geographical, and nutritional—within which a species can exist, and they distinguish a preferred niche, where the conditions most favor the species' thriving, and a realized niche, the conditions with which the species has in fact to make do.

Behavior can constitute an important part of adaptation to a particular niche as in specialized ways of obtaining and dealing with specific foods.

Ecology The study of the relationships between organisms and their physical and biological environments. It is subdivided in various ways: according to the kind of habitat (thus the ecology of lakes is part of limnology, and marine ecology is part of oceanography); according to the kind of organism (plant ecology, animal ecology); and according to whether the focus is on individual species (autecology) or the complexes of organisms constituting communities (synecology). Behavioral ecology considers species-typical behavior in terms of ecological considerations and is thus part of sociobiology. Also overlapping with ecology is population biology, which deals with the genetics and demography of populations.

Ecotype Adaptation type. Ecotypes are animal (and plant) populations characterized by their adaptations to specific environmental conditions and thereby distinguishable from other populations of the same species. The most common differences between ecotypes are in physiological characteristics, such as responsiveness to environmental factors governing annual periodicity. There may also be morphological variation between ecotypes, especially in plants, and animals may show behavioral variation, such as differences in food or habitat preferences. Imprinting processes can be involved in the establishment of ecotype differences (see FOOD IMPRINTING; HABITAT IMPRINTING). For the establishment and maintenance of ecotypes a degree of reproductive isolation is necessary, and consequently in many cases distinctive recognition marks are developed.

83

Ectoparasite   See Parasitism

Effector   Executive organ. A general term for muscle and gland cells and the muscles and glands they constitute; also color cells that are innervated and hence under the direct control of the central nervous system. The nerves running from the central nervous sytem to effector organs are described as efferent.

Efference   All the nervous excitation that goes from the central nervous system to the effectors. The transmitting nerves are called efferent or motor nerves.

Efference copy   See Reafference

Egg follicle   See Ovulation

Egg rolling   Egg retrieval. A movement sequence performed by many ground-nesting birds when an egg is found outside the nest, to roll the egg back into the nest cup. From the direction of the nest, the bird stretches its neck out toward the egg, places the underside of its bill against the far side of the egg, then bends its neck to roll the egg toward its breast. In some species the neck bending may be accompanied by stepping backward, particularly when the egg cannot be reached from inside the nest. The bird counteracts deviations in the travel of the egg by side-to-side adjustments of the angle of the bill. The egg-rolling movement is of historical interest because it provided the first example, in the greylag goose, of what was recognized as a compound behavior pattern consisting of a fixed action pattern and a taxis component. If the egg is removed after a bird has begun the egg-rolling movement, it will complete the bending movement in the vertical plane, but cease the sideways correcting movements. The egg rolling itself is thus a fixed pattern, which, once begun, runs to completion without external stimulation. The balancing movements, on the other hand, are oriented reactions, whose occurrence and course depend on the prevailing stimulus situation—in this case the moment-to-moment situation of the egg.

Ejaculation   See Copulation

Electrocommunication The capacity of "electric" fish to transmit social signals by means of electrical pulses. Generation of electrical pulses for this purpose has evolved independently in two groups of fish: the mormyrids of Africa and the gymnotids of South America. Originally it served for the detection of obstacles and prey under conditions, such as turbidity, where the other senses would presumably not suffice, allowing the fish to register distortions caused by such objects in the pattern of self-generated electric fields. From this function electrocommunication has been turned to social ends in some species, for territorial advertisement, group cohesion, and possibly other functions.

Electroencephalogram (EEG) See Arousal

Emancipation Disengagement during ritualization of derived activities from their original causal control (for example, as displacement activity) and their transferral to the control of the causal system (for example, courtship motivation) in which they function as communication. Emancipation is a form of motivational change (compare CHANGE OF FUNCTION).

Embryo The developmental stage extending from fertilization of the egg to birth or hatching. Especially toward the latter end of this stage, embryos can make distinct and controlled movements, including, in some birds, vocalization (see REFLEX; SPONTANEOUS BEHAVIOR). Some birds have been credited with learning prior to hatching (see PRENATAL LEARNING).

Embryonic development See Ontogenesis

Emotion For animals, as for people, this word is used with a variety of meanings, including inner state of feeling; physiological state of arousal as indicated by heart rate, skin conductance, or EEG; and action tendency as indicated by such outward signs as facial expression, tone of voice, or body posture. Partly through the influence of behaviorism, ethologists and comparative psychologists have generally suppressed the subjective connotations of "emotion" in connection with animals. If they use the word at all it is likely to be in a sense close to that of motivational state or arousal or action tendency. The same holds for words referring to kinds of emotion, such as anxiety, fear, hostility, and sexual excitement. The open-field test is a standard method used by experimental psychologists for measuring what they call "general emotionality." Darwin pointed the way for ethology by emphasizing the expressive aspect of emotion. However, he also took for granted that animals experience emotions as subjective states, and this point of view has recently been revived by cognitive ethology in its concern with the question of animal awareness.

End act Consummatory act. A fixed action pattern that comes at the end of a series of appetitive actions (see APPETITIVE BEHAVIOR). According to the classical ethological theory of motivation, the end act consumes the energy propelling the sequence and so leads to discharge or dissipation of the drive (compare DRIVE REDUCTION). (The linking of "consuming" and "consummatory" was an unintended pun by the classical ethologists. The two words have different origins: "consuming" comes from Latin *consumere,* meaning to devour or destroy; "consummatory" comes from Latin *summa,* a total or sum.) More recent evidence indicates that the stimulation consequent on performance of the end act is responsible for the cessation of the appetitive behavior. Hence the termination of such sequences is often attributed to consummatory stimulation. However, even the distinction between appetitive behavior and end act has been called into question, since at least some kinds of appetitive behavior can reduce the likelihood of immediate repetition or of continuation of the sequence in which they occur (see ACTION POTENTIAL). Examples of end acts include movements of food ingestion (chewing, swallowing) in the functional system of feeding and copulation movements in sequences of sexual behavior. An end act may be preceded by a long series of appetitive acts under certain conditions, as in the finding and capture of prey or the mutual stimulation of sex partners to achieve the synchronization necessary for effective mating.

Endocrine glands Ductless glands. Glands that synthesize and secrete hormones into the bloodstream.

Endogenous From inside to outside; pertaining to causation or control located or originating within the body, central as opposed to peripheral. In behavioral science the term is sometimes regarded as tendentious, because some people mistakenly equate it with innate. However, "endogenous" means only that an overt or externally measurable characteristic, such as a behavior pattern or a biological rhythm, originates within the organism. Nothing is implied about the genetic basis of this characteristic, since internal factors can be affected by experience or some other impress of earlier environmental influence. The term is particularly useful in discussion of biological rhythms, where it refers to the inner basis of cyclical changes in many biological phenomena (see PERIODICITY). In ethology it includes spontaneity; in psychology it is more or less synonymous with "intrinsic."

Endogenous rhythm See Periodicity

Endoreceptor See Interoceptor

Energy model See Motivation

Engram Memory trace; the physical basis of learning. A material or structural change in the cells or cellular connections of the central nervous system consequent on patterns of neural activity resulting from experience. Various theories have been presented about the nature of engrams, ranging from neural circuits (see CELL ASSEMBLY) to molecular configurations in cells to patterns of force fields. Solving this problem is high on the agenda of those investigating the physiological basis of memory.

Entrainment The synchronization of a biological rhythm with an environmental rhythm (see ZEITGEBER).

Environment This term has various uses in biology and psychology. In its widest sense environment signifies the totality of factors constituting the outer world of an organism, without distinguishing between those factors that affect the organism and those that do not. More often, however, the distinction is drawn between factors relevant and irrelevant to the organism, especially when the German equivalent—*umwelt*—is used, as it frequently is in ethological literature, even in English. The term then refers only to the totality of those factors of the surroundings that actually exert some influence on the organism or that are affected by the organism, such as climate, food supply, predators, and competition. In this sense the environment can be very different for different organisms occupying the same habitat.

An animal's psychological or behavioral environment is determined by its sensory capacities and their "tuning" (see STIMULUS FILTERING), and also by the ways in which it can act on or in relation to its surroundings, as in locomotion, for example. People who believe that all or most significant behavior (or knowledge, if epistemology is the issue) is a product of experience are sometimes referred to as environmentalists; this does not imply that they are concerned about nature conservation.

Enzyme   Organic catalyst. A substance that facilitates specific chemical reactions in the body without being more than transiently changed in the process. The body has a great variety of enzymes, which are of crucial importance for all phases of metabolism, both intracellular and extracellular: such as respiration, digestion, and excretion. They are also vital to the action of genes, a high proportion of which encode recipes for the synthesis of enzymes: hence the slogan "one gene, one enzyme."

Epideictic display   See Group selection

Epigamic display   Courtship behavior performed in a situation of epigamic selection, that is, where individuals of one sex compete with one another through display to be chosen as mates by members of the other sex. Usually it is the males who perform epigamic display, as in the lek behavior of ruffs and sage grouse. See SEXUAL SELECTION.

Epigamic selection   See Sexual selection

Epigenesis   Biological development viewed as a process in which each stage in the growth and differentiation of an individual, from zygote to mature adult, arises from the preceding stage through the joint action of genetic and environmental determinants. In this view, at any point in development the phenotype already constructed sets the stage, as it were, for further construction, guided by the genes now brought into operation and environmental influences. Thus phenotype, genotype, and environment work in dynamic conjunction and progressive sequence. Epigenesis contrasts with preformationism, according to which the organism is, so to speak, prepackaged in miniature or in coded information in the zygote, which has only to be caused to grow for normal development to proceed. Modern views that liken the genetic program to a blueprint and consider the environment as supplying no more than the materials required for development are preformationistic. Classical ethology tended toward preformationism; modern ethology is more inclined toward epigenesis.

Epiphenomenon   In English this word usually connotes an effect that is causally inconsequential and incidental or secondary to one that is operative or primary. For example, the squeaking sound that emanates from the turning of a pulley does not contribute anything to the work accomplished by the pulley. Mechanistic materialists have argued that mentality is an epiphenomenon and hence that we are deluded when we think, for example, that willing our actions causes them, since the only causes at work are physical ones, which produce mental states but are not affected by them.

German ethologists use the word *Epiphänomen* to refer to an overt concomitant of a physiological state or process. For example, arousal or activation of the autonomic nervous system in situations of danger or threat can cause urination, defecation, trembling, feather ruffling, or hair erection. Such symptomatic expression has particular significance for ethology when it indicates the evolutionary origin of display behavior. The signal function of the behavior may be enhanced through ritualization, and sometimes morphological support evolves. For example, many animals show vigorous tail wagging when aroused. In some species these movements have become threat displays, and in rattlesnakes and porcupines the tail waving also produces an acoustic signal, as a result of modification of structures (vertebrae in snakes, quills in porcupines). Also the excretion of urine for territorial marking is probably derived from an epiphenomenon—urination caused by "anxiety." The same applies to the use of feces by hippopotamuses, for example, to indicate territorial boundaries. The epiphenomenal sounds produced by breathing may constitute the evolutionary antecedent of vocalization in many cases.

Epitreptic behavior   See Apotreptic behavior

Erroneous imprinting   Malimprinting; fixation of a behavior pattern on something other than the natural object as a consequence of faulty experience during the revelant sensitive period. Under free-living conditions, animals rarely experience erroneous imprinting. It has occasionally been reported of ducks and geese, most often when a female whose laying has been disturbed raises the brood of another species laid in her nest. Songbirds also have presented the odd case; for example, American white-crowned sparrows overwintering in the same region as other subspecies with different dialects have been known to acquire the wrong dialect (see SONG IMPRINTING). Erroneous imprinting can be induced experimentally, as when a young animal of one species is raised by a member of another species. This can affect all forms of imprinting, (such as the following response and sexual imprinting). The effects of erroneous imprinting may persist for a long period, and even for life (see IRREVERSIBILITY). Erroneous imprinting of hand-reared animals to their caretakers is sometimes referred to as human imprinting.

Escape behavior   Fleeing. Flight from or evasion of a situation of danger or threat. Examples include an animal's putting distance between itself and a predator or rapidly retreating from the aggressive display of a conspecific; also withdrawal into burrows or shells, submergence, protean behavior, death feigning, hiding, and other forms of concealment, perhaps using some form of camouflage. Escape is a form of defensive behavior.

Estivation   See Hibernation

Estrodiol   See Estrogens

Estrogens   A group of female sex hormones that in vertebrates are produced chiefly by the ovary. They are involved in egg development and in development and maintenance of the female sex characteristics and reproductive behavior. Like the other sex hormones, estrogens are steroids derived from cholesterol. The similar molecular structure of these substances makes it relatively easy for them to be converted into one another. There is evidence that at least some of the effects of the male sex hormone testosterone (see ANDROGEN) result from its conversion to estrogen within the cells concerned. The most familiar estrogen is the follicle-secreted hormone estradiol.

Estrus   Heat. In mammals, that phase of the reproductive cycle when the female is sexually receptive and copulation can effect fertilization (see OVULATION). This change is primarily physiological, to provide for pregnancy. But it also causes behavioral changes (behavioral estrus or estrus behavior), such as soliciting and other patterns connected with copulation (for example, lordosis). In many primates the female's appearance also changes, as in the conspicuous swelling and coloring of the genital region in many species. These changes, which constitute primary sexual releasers for the males, may also lead to altered positions within the group; males take to staying close to estrus females, and so consort pairs are formed. To a large degree estrus phenomena are governed by hormones (see ESTROGENS; PROGESTINS). In contrast to males, which display continuous sexual appetite during the reproductive season, female mammals are more or less sexually inactive outside of the periods of estrus (see ANESTRUS).

Ethoendocrinology   See Behavioral endocrinology

Ethogram   Behavioral inventory; catalogue of actions; a survey, as complete and precise as possible, of all the behavior patterns characteristic of a species. Compilation of an ethogram provides a starting point for experimental investigation of species-typical behavior and for interpretation of the results. A descriptive behavior study or descriptive ethology has to contend with the critical problems of categorizing and labeling behavior patterns, and, if called for, sorting them into principled classes. Also, analysis of a longer behavior sequence may reveal temporal associations among its components (see SEQUENCE ANALYSIS).

In the early years of animal behavior study, ethograms were compiled mainly from what ethologists called observation protocols. Nowadays numerous aids are available, which help in the making of protocols, and allow for precise analyses and permanent records. Thus observations can be spoken into a tape recorder tape, avoiding the interruption entailed in writing down notes. Through the use of film and video and audio tape-recording, movements and sound sequences can be recorded and analyzed into their constituents, compared with one another, and quantitatively investigated, for example, through frame-by-frame analysis. To analyze complex movements, some ethologists have used the notational methods of choreography to describe the motions of a single limb or its parts in relation to the rest of the body parts, to the animal's surroundings, or to the behavior of other individuals.

Ethological isolation   See Reproductive isolation

Ethology   The biological study of animal behavior. Ethology comprises study of the immediate (proximate) causes of behavior, including external stimulation and internal physiological mechanisms and states; development, including the ontogenesis and genetics of behavior; biological function, including immediate effects and ultimate consequences (see SURVIVAL VALUE); and evolutionary origins and modifications (see PHYLOGENY). Conceived in this way, ethology embraces an enormous range of investigation from biochemistry to sociology and anthropology, sharing interests with various branches of physiology, genetics, ecology, psychology, anthropology, and even linguistics and philosophy. Consequently its boundaries with respect to other sciences of behavior are rather ill-defined. This was not always so. The historical roots of what has been described as the "hard core" of ethology can be traced to the comparative morphological tradition that used to dominate zoology, the avocational pursuit by amateurs of natural history, and the evolutionary interpretation that Darwinism brought to these studies. As applied to a school of animal behavior study, the term "ethology" was first adopted by a small group of European zoologists, headed by Konrad Lorenz, which emerged shortly before the beginning of the Second World War. This group emphasized the consideration of behavior in relation to an animal's natural way of life and environment, hence the adaptive significance and evolution of behavior, hence the genetic inheritance of behavior and the basis it gave for a theory of instinct. The group came to be associated especially with the theory of instinct, largely as a consequence of the success of Tinbergen's (1951) book *The Study of Instinct.* This phase is sometimes referred to as classical ethology. For some people the connotation of the term is still much as it was in 1953, when criticisms of instinct theorizing began to appear, both from within ethology and from other schools. Since then the theoretical unity of the original group has been lost, as ethology has become divided among different special interests and has tended to merge with neighboring disciplines. Hence the present situation of wide range, baffling diversity, and fuzzy outlines of "ethology." The historically narrower and currently broader senses of the term can cause confusion if context does not clarify which is intended.

Ethology of domestic animals   See Applied ethology

Ethometry   See Quantitative ethology

Ethomimicry   See Behavioral mimicry

Eusocial   Along with "semisocial," a term applied to social insects which have reproductive division of labor with a more or less sterile caste of individuals that take care of the brood. The distinction between the two terms turns on the durability of the social group and the genetic relatedness of its members. Semisocial species are those in which the generations do not overlap, and hence all the cooperating individuals are sisters. They include bumblebees, the socially primitive bees, and wasps. Eusocial species are those in which two or more generations overlap, with consequent attenuation of kinship between helper and helped. The eusocial species comprise all termites and ants and the socially most complex bees and wasps. A mammalian parallel to the eusociality of social insects has been found in mole rats (*Heterocephalus glaber*).

Evolution   In modern biology, the changes undergone by organisms in the course of their phylogenetic histories (for organic evolution, see PHYLOGENY). More strictly speaking evolution consists in changes from one generation to the next in gene frequencies of the gene pools of populations. These changes generally lead to the

93

emergence of complex organizations and specialized forms from simpler precursors. The most important agencies of organic evolution are mutation, which provides variation, and selection, which determines the variations that are preserved and perpetuated. Behavior is subject to evolution in this sense just as all other characteristics of an organism are. Indeed, behavior may sometimes serve a kind of pacemaker function in that behavioral change (for example, the development of a new food preference or mode of foraging) can be a first step of evolutionary adaptation, followed by the natural selection of appropriate changes in body form and other characteristics.

In organic chemistry combination processes have been discovered that follow a course of progressive development in a manner analogous to organic evolution; consequently such processes have been described as constituting chemical evolution.

A third application of the term is to cases of a new behavior pattern arising in a population and being passed on to subsequent generations by tradition, described as cultural evolution. In contrast to the other two forms, cultural evolution can be waywardly variable and even show reverses and so has the greatest degree of openness (see PROGRAM). It is exceptional also in that it can be construed as a kind of Lamarckian process because in a sense it involves inheritance of acquired characteristics.

In the nineteenth century Herbert Spencer developed a concept of general evolution, which applied to progressive change in the cosmos, the geological record, society and culture, mentality, and individual development, in addition to chemical and organic nature. Although this synthetic philosophy had a number of important influences, Spencer's notion of general evolution is now regarded as muddled—a confusion of several distinct processes that must be considered apart from one another.

Evolutionary stable strategy (ESS)   When two or more alternative ways of coping with a situation vie with one another, either in a population or in an individual's options, and each alternative is best in some circumstances but not in all, the proportions in which the alternatives occur may reach an equilibrium determined by the combination that yields the highest net return in, for example, reproductive success. In a population in which reciprocal altruism has

94

evolved, individuals that accept help without ever returning it ("spongers") will initially have an advantage over helpers because they will enjoy the benefits of the altruistic system without incurring any of the costs. Spongers will therefore be selected: they will increase in the population at the expense of the helpers. But as the proportion of helpers decreases, sponging becomes less and less profitable. At some point the advantage the remaining helpers get from one another outweighs the disadvantage of being exploited, and they make a comeback. This will be checked as their recovery provides an advantage to the spongers. The proportions of helpers and spongers will eventually converge to an equilibrium ratio at which net reproductive success is maximal; that ratio of spongers to helpers becomes the evolutionary stable strategy. Any shift away from it results in decreasing success for the increasing moiety. This argument, derived from mathematical games theory, has been applied to such matters as the evolution of threat display and parental investment. It was originally deployed in evolutionary biology and ethology by Maynard Smith (1974).

Evolutionary vestige   See Relict

Exafference   Afference resulting from stimulation of exteroceptors through environmental change with respect to which the animal is passive; for example, a cloud passing in front of the sun will cause the animal to perceive a drop in light intensity not brought about by its own activity. Compare REAFFERENCE.

Experience deprivation   See Isolation experiment

Exploratory behavior   Curiosity behavior; exploration. Search for and active investigation of novel situations in the absence of pressing need. It is most prevalent in higher vertebrates—birds and mammals—but has been sporadically reported for fish and reptiles as well. In general, exploratory behavior is most conspicuous in the young, yet in several groups (raptors, parrots, rodents, and primates), it persists throughout life.

Exploratory behavior has much in common with play behavior; indeed no sharp line can be drawn between the two. Thus, just as many animals have a compelling need to play, so do they appear to have a distinct appetite for novelty or curiosity drive (see APPETITIVE BEHAVIOR); from time to time they appear to spontaneously seek out new stimulation and objects to investigate. Like play, exploratory behavior occurs only at times when no other behavioral tendencies are activated, and it consists to a high degree of free combinations of different kinds of behavior, often in novel sequences. In this it differs from trial-and-error learning, which draws on behavior from within a single functional system, such as foraging, or nest building. The function of exploratory behavior appears to be similar to that attributed to play: it provides opportunities for gathering and refining the individual's knowledge of objects and space and for shaping movement patterns to the characteristics of those objects.

The terms "exploratory behavior" and "curiosity behavior" are used more or less synonymously in the animal behavior literature. Occasionally, however, a writer will distinguish between exploration, referring to investigation of new spatial circumstances, and curiosity, referring to investigation of novel objects. Many authors completely avoid the word "curiosity" because it can so easily be taken to imply the existence of consciousness.

Expressive behavior   See Display

Extension of function   The acquisition of new or additional function by some feature of an organism in the course of its evolution. A behavioral example is courtship feeding: what was originally a means of provisioning a female with food to help meet the extra demands of producing offspring has evolved in many cases into behavior that secondarily contributes to strengthening of the pair bond and perhaps provides the female with a way to assess the courting males as potential providers for the brood she will have.

External stimulation   Sensory stimuli from outside the body. It is sometimes also referred to as peripheral stimulation. See EXTEROCEPTOR.

Exteroceptor   Sense cells or sense organs that, in contrast to interoceptors, register external stimulation. The classical five senses of vision, hearing, smell, taste, and touch are all served by exteroceptors. See RECEPTOR.

Extinction　In animal behavior study this word has two distinct meanings. In the context of evolution it refers to the dying out of a species. In the context of learning experiments it means the loss of a habit through discontinuation of reinforcement. Extinction in this latter sense is not to be equated with forgetting, for the response or association can be reinstated immediately if the reinforcement is restored. Moreover, forgetting is not a consequence of absence of reward, but usually a concomitant of the passage of time. As a rule, extinction results in the disappearance of behavior much more quickly than "mere" forgetting.

Extirpation　Cutting out; the removal of an organ or part of an organ. In behavior study extirpation usually refers to experimental removal of brain areas to investigate localization of function. This quite crude approach has been largely replaced by more refined methods, such as the making of small lesions or electrical stimulation though microelectrodes (see BRAIN STIMULATION).

Extrinsic selection　See Selection

FOOD BEGGING

F

Facial expression Expressive movements or configuration of muscular contraction in the region of the face. Facial expressions serve a communication function and are especially pronounced in contexts of agonistic behavior. A facial expression can convey an animal's "mood" or behavioral tendency (see ACTION POTENTIAL; MOTIVATION) and can arouse or change the behavioral tendencies of individuals toward which it is directed.

Facial expression occurs only in the more highly evolved mammals, which have the requisite facial musculature and innervation. Comparatively simple forms are found in, for example, canine and feline carnivores and many ungulates (see FLEHMEN); many primates, most notably chimpanzees, have an almost human richness of form of expressive facial mobility.

Facilitation A term with several different applications in physiology, psychology, and ethology. In physiology it refers to the additive effect that multiple nerve impulses arriving at a synapse may have in bringing about synaptic transmission. Temporal facilitation is the summation of results from impulses arriving close together in time via a single nerve fiber; spatial facilitation occurs when the summating impulses arrive via different nerve fibers more or less simultaneously. The terms "temporal summation" and "spatial summation" are interchangeable with "temporal facilitation" and "spatial facilitation" (see SUMMATION).

In psychology facilitation can refer to enhancement of behavior (for example, quickening of perception or response) as a consequence of contributory internal or external factors or processes, such as attention, arousal, or reinforcement. In ethology, somewhat similarly, the term sometimes refers to the accelerating influence of environmental factors on behavioral development during the course of ontogeny (as distinct from induction, which refers to the calling forth of new features). However, ethologists more often use the term to mean social or behavioral facilitation: the inducing of an individual to begin or resume an activity because others nearby are engaged in it. Thus a satiated chicken can be made to continue feeding by placing it in the company of hungry chickens.

Facing away   See Appeasement behavior

Factor analysis   A statistical method for analyzing the degrees of covariation among a number of different measures of a capacity such as intelligence or of a type of behavior such as courtship. Factor analysis involves working out the partial correlation coefficients between the different combinations of variables and treating them in such a way as to reveal dependence on a smaller number of common factors. Ethologists have used this and similar techniques (hierarchical cluster analysis, multivariate analysis, principal component analysis) to generate evidence for the existence and organization of drives or motivational systems underlying behavioral patterning.

Fall song   See Subsong

Family   This term is used in two different areas of biology. In taxonomy it is a level of classification that groups together closely related genera. (Families in turn are grouped into orders at the next higher level of taxonomy.) For example, the catlike carnivores belong to the family Felidae, the doglike carnivores constitute the Canidae, and the weasellike carnivores, the Mustelidae.

FAMILY

In ethology a family is a social unit consisting of a kin group: one or both parents living with their offspring. The main function of the family in this sense is provision of parental care. Families occur in all mammalian species, in almost all birds, in some other vertebrates, especially fish, and among some invertebrates, for example, insects, spiders, and myriapods.

According to the parents' share in care of the young, families may be two-parent, maternal, or paternal. Two-parent families in which both parents are equally or similarly concerned with the young are found in the majority of birds, some fishes (many cichlids), and some mammals (gibbons, canids). In other cases the parents' contributions to care of the young differ (see DIVISION OF LABOR). In many birds the male takes no part in incubation but contributes more or less equally with the female in providing food for the nestlings. The extreme of this division of labor is one parent taking sole charge. The maternal family is typical of mammals, where the mother's parental role is obligatory because she alone produces the milk. It occurs also in many gallinaceous birds and hummingbirds, fishes, spiders, and scorpions. Paternal families are found only occasionally, as in sticklebacks and labyrinth fish, and, among birds, in phalaropes, painted snipe, and rheas.

The societies of insects also take the form of families: maternal in the Hymenoptera (ants, bees, and wasps) and two-parent in the termites, in which the males remain with the queen after the nuptial flight and contribute to the upkeep of the colony.

As a rule the animal family breaks up after the young become independent and is consequently only a temporary social unit. However, in many species with two-parent families the two adults may remain together as a pair. In some species the young stay with both parents or with the mother beyond the time of independent foraging, so the family may include young of different ages. This extension of family life is found in species where the young have to acquire essential information, such as knowledge of the route to winter quarters for migratory birds, through tradition, which is possible only after a certain age, and in species in which the young enter into a group formation with the parent(s) beyond the period of their own rearing. In such circumstances the older young may contribute to the care of their younger siblings (see HELPER). Among mammals, for example, gibbons and many ungulates, persisting or long-lasting families are quite common.

Fatigue   In ethology the term usually refers to specific exhaustibility: the reduced releasability of a behavior pattern as a consequence of preceding performance (see THRESHOLD CHANGE). If this
reduction is caused by exposure to the releasing stimulus, the
change is described as stimulus-specific waning of response (see
HABITUATION); if the refractoriness is caused by performance of the
behavior itself, it is described as action-specific exhaustion. Neither
form has anything to do with the idea of tiredness or weariness
connoted by "fatigue" in other contexts.

Fear   When ethologists apply this word to animals they usually
leave aside the usual subjective connotation: at most they may refer
to an emotion as manifest in outward signs or behavioral tendencies;
more often they mean a motivational state that impels fleeing, escape, or defensive behavior in a situation of danger or threat (see
ESCAPE BEHAVIOR). Fear can thus be a component of stress or of
conflict situations in which fleeing is opposed by sexual attraction,
family ties, an urge to attack, or some other motivation. Such a state
of conflict can give rise to displacement activity, ambivalent behavior,
or redirection. Fear aroused during social interactions may be conveyed by signals, such as facial expressions in most primates, and
may cause the fearful animal to show appeasement behavior. Fear
reactions to predators include the giving of alarm calls or other
warning behavior. An animal's experience of fear in a particular
situation can condition it to avoid that situtation (see AVOIDANCE
CONDITIONING).

Feather ruffling   Erection of feathers. Without doubt the primary
function of feather ruffling is temperature regulation. In this connection three feather positions have been distinguished: ruffling, or
extreme erection, with feather tips separated from one another, thus
opening a passage for air to get to the skin for cooling; fluffing, in
which erection stops short of separation of the feather tips and so
increases the thickness of the insulating layer for heat retention;
and sleeking, which contrasts with erection in that the feathers are
depressed against the skin surface, thus reducing the thickness of
insulation and reducing heat retention. Feather erection may be
restricted to certain parts of the body, depending upon the conditions.
For example, the feathers around the brood patches are ruffled as

part of the sequence of settling movements with which a bird lowers itself onto its eggs to incubate them.

Special forms of feather ruffling have frequently evolved for the secondary function of display, often with associated modification of the shape or coloration of particular feathers, as in crests and plumes, the capes of pheasants, and the ruffs of ruffs. Such ritualized ruffling is especially prominent in the displays of threat behavior and courtship, where they may function as releasers necessary for the conduct of social interaction. For example, in many of the species in which allopreening occurs, a slight ruffling of the plumage can indicate where on its body the recipient of the attention would like the preening directed (see SOCIAL GROOMING).

Feedback   When the output of a system has consequences that the system registers in such a way as to influence its further functioning, it is said to be regulated by feedback. A familiar example is the automatic oven: the thermostat registers the temperature of the stove, turning the electricity off when the temperature reaches a preset level and turning it on again when the temperature drops below that level. Control of body temperature is effected in analogous ways, as is control of blood pressure, blood sugar level, and the processes of homeostasis in general. These are examples of negative feedback: the mechanism negates deviations away from a set value, and the output of the system eliminates its cause. Negative feedback is a means of conserving conditions, maintaining stability. Positive feedback obtains when the output of the system adds to its cause, resulting in progressive buildup of output. An example is the squeal that occurs when a microphone is brought near a connecting loudspeaker: the sound from the speaker is picked up by the microphone and returned through the system, thus adding to the speaker output in augmenting cycles. A physiological example is the generation of a nerve impulse: local depolarization of a neuron membrane increases the membrane's permeability to sodium ions at that point; the consequent influx of sodium ions further depolarizes the membrane, further increasing sodium influx, and so on, until the limiting conditions are reached and restorative processes can take over. As this example illustrates, positive feedback can serve positive ends in the functioning of the body but because it carries a system away from a stable state, as a transient occurrence contained by extra-

102

neous limiting conditions, it is a much less common feature of control systems than negative feedback. Another distinction sometimes encountered is that between "internal" feedback, when the feedback loop is contained within the system, and external feedback, when the loop extends outside the system. However, whether the feedback is considered internal or external depends on what the system is conceived as including. If a motor neuron is the system, the negative feedback conveyed by its Renshaw cell is external; but if the system is the whole motor control mechanism, then the Renshaw circuit is internal to it.

A notable use of the feedback concept in ethology is the reafference principle (see also TOTE). Both negative and positive feedback, and internal and external feedback, have numerous examples in the area of hormones and behavior and in the application of control theory (see CYBERNETICS) to questions of motivation.

Feed-forward "Change that anticpates a need" (Barnett 1981). Animals often act on the basis of cues that signify what condition is to be expected, rather than just on the basis of current physiological state and feedback. For example, doves or rats made thirsty to the same degree drink different amounts of water depending upon the temperature, thus anticipating the dehydrating effects of hot weather. Similarly, many animals drink prior to eating and thus anticipate the dehydrating effects of digestion.

The concept also has application in neurophysiology. For example, when the stretching of a limb muscle elicits reflex contraction of that muscle, feed-forward inhibition prevents contraction of the antagonistic muscle: in addition to its excitatory connection to the motor neuron running to the stretched muscle, the afferent neuron from that muscle synapses with an inhibitory interneuron running to a motor neuron of the antagonistic muscle. Similarly, the efference copy postulated by the reafference principle, or the corollary discharge postulated by Teuber (1960), can be considered feed-forward preparation for the stimulus changes consequent on action.

Feeding territory See Territory

Fertilization Union of the nuclei of egg and sperm. Fertilization and insemination are not sharply differentiated in the literature, often being used interchangeably. Fertilization may sometimes have the same meaning as "mating" (see COPULATION).

Fidelity to place   Site tenacity; similar to and associated with homing. The propensity of many animals to return to the place of birth, the place of first breeding, or any other previously occupied locality, such as winter quarters for migratory birds. Notable place fidelity is found in some bats, many migratory birds and fish (salmon and eels), and a few mammals (gray whales, *Eschrichtius*). Fidelity to place can be an important factor in maintaining reproductive isolation. In birds early learning is usually the basis of fidelity to the place where they were hatched (place imprinting). The German word *Ortstreue* is sometimes used by English writers.

Fight and flight syndrome   See Stress

Fighting   See Injurious fighting, Ritualized fighting

Filial imprinting   Following-response imprinting. The rapid learning by which newly hatched young of precocial birds have their filial responsiveness fixated on or tuned to the characteristics of one or both parents. The chicks react to the first moving object they see, normally one of their parents, by following it, and the following response is tied to the characteristics of that object from then on. Consequently it is possible experimentally to imprint the following response of such birds on all sorts of things—wooden boxes, balloons, even people and flashing lights. Following-response imprinting is the most extensively studied of all forms of imprinting. As a rule it is restricted to an extremely brief sensitive period. However, more recent study has shown that there may be more flexibility than was originally supposed in the timing of the sensitive period and in the extent to which the imprinting is irreversible. Also, contrary to earlier ideas, reinforcement may be involved.

Final common path   "The final or efferent neurone is, so to say, a public path, common to impulses arising at any of many sources of reception" (Sherrington 1906). The pool of motor neurons controlling a particular movement can be entered via a variety of pathways from local reflex connections or higher control areas in the central nervous system. The movement can thus be included in a variety of reflexes and actions. A dog may wag its tail in response to an insect bite or as an expression of excitement or to avoid having it stepped on. In each case the different controlling paths converge upon the same motor neurons and muscles. Ethologists sometimes write about competition for the final common path when an animal is subject to conflicting motivation, meaning that alternative courses of action are vying for control of the motor system.

Fitness   In current biology, reproductive success of a genetic entity as gauged by its proportional representation from one generation to another. Although fitness can apply to single genes in competition with one another or to competing groups of organisms, the more common reference, at least in contexts involving behavior, is to the reproductive success of individuals and hence the perpetuation of their genotypes. In this usage a distinction is drawn between individual and inclusive fitness. Individual fitness is the success of an individual in leaving copies of its genotype in the next generation relative to that of other individuals with their genotypes. It is therefore a function of the variety of genotypes in contention (if there were only one genotype, its fitness would be absolute) and of the conditions affecting the odds among them. In the classic example of the peppered moth, *Biston betularia,* genotypes for dark-colored individuals are more successful (fitter) in industrial regions than those for light individuals, because the dark moths are better camouflaged against soot-begrimed tree bark and hence less subject to predation than the light moths; in rural areas the reverse is the case because clean bark favors the light moths.

Inclusive fitness takes into account the extent of an individual's efforts to increase the representation of its relatives' genes in the gene pool, over and above what it adds through its own reproduction. The more closely two individuals are related genetically, the more genes they have in common, and hence the more each has to gain, in terms of genetic representation in the gene pool, by furthering the other's individual fitness. In special cases such gain can be greater than that from reproduction. In beehives, because of the peculiar mode of sex determination (see haplodiploidy), the females produced by the queen have more genes in common with one another (75 percent) than they would have with their own daughters

(50 percent), so the fact that most are sterile workers supporting a fertile sister or two is consistent with notions of fitness. Behavior toward relatives that conduces to increase in inclusive fitness is favored by a process called kin selection.

The use of the term "fitness" for the modern biological concept derives from Herbert Spencer, who introduced it in the phrase "survival of the fittest." However, the nineteenth-century connotation of being favored physically in the "struggle for existence" has only a faint echo in the essentially statistical nature of current evolutionary thought.

Fixed action pattern A stereotyped pattern of movement that may be species-characteristic; spatiotemporal sequences of muscle contractions that are relatively constant in form and that generally belong to a functional system. Such movements can occur either alone (preening and shaking movements) or as constituents of a behavioral complex, such as behavior patterns of courtship. Fixed action patterns are elicited by external stimulation of specific sorts, which may also affect their intensity and orientation (see TAXIS COMPONENT). However, the morphology of such a movement—its spatiotemporal ordering—is largely or completely independent of external or peripheral stimulation.

When endogenous motor patterning of this sort is a species-specific characteristic, it can be argued that it is based in a genetic program. This was generally assumed to be the case in the early theorizing of ethology, from which the term "fixed action pattern" derives. Lorenz (1937) regarded a special class of fixed action patterns as *sui generis*—the only patterns of behavior that possessed all the attributes of instinct and hence could be called instinctive activities (*Instinkthandlungen*). In addition to being stereotyped and species-specific, these fixed patterns terminate bouts of appetitive behavior, which are aimed at finding the stimuli necessary for their elicitation. On this basis it was proposed that the appetitive behavior is caused by accumulation of energy specific for each kind of instinctive activity (see ACTION POTENTIAL), the energy being expended in performance of the fixed pattern. With the abandonment of this theory of motivational energy and the realization that stereotyped patterns are rarely as independent of environmental factors as was originally thought, the term "modal action pattern" is now frequently used instead of "fixed action pattern" to avoid the older connotations.

Although the ontogenetic basis and causal control of stereotyped action patterns are now more open to question than they were for classical ethology, there is no question that they occur. Movement patterns that have constant form can present taxonomic characteristics as good as or better than morphological patterns.

Flight distance   The distance to which an animal will flee when alarmed by a predator, say, or when intimidated by a threat from a rival. Flight distance varies from species to species, as well as within a species according to an individual's experience or social status and the nature of the situation. The flight distance of hand-raised animals may be very short. Also in zoological gardens, parks, and even nature reserves it is often far less than in completely wild conditions.

Flehmen   Lip curl. A facial expression shown by many mammals, especially ungulates and felid carnivores, less so by insectivores, bats, and primates. The upper lip is raised or pulled back, exposing the upper incisors or gum ridge. Often it is accompanied by a tilting upward of the head, forward extension of the neck, and closing of the nostrils. Flehmen is usually associated with sniffing, as when males nose the urine of females.

The functon of flehmen has yet to be conclusively established. It probably serves to convey odorous substances in solution to the vomeronasal organ (Jacobson's organ), an olfactory receptor area in the roof of the mouth. In lower vertebrates this structure is the means of sensing the smell of food in the mouth, but in macrosomatic mammals the vomeronasal organ is apparently specially tuned to female sex hormones. The concentration of these hormones in the female's urine fluctuates with her reproductive cycle, so a male can gauge the female's estrus status by using the lip movement to sample the smell of her urine.

Flock   A form of open assemblage or concentration of birds or mammals (see GROUP FORMATION, HERD). Comparable grouping in crustaceans (for example, water fleas) and insects (locusts, midges) is described as a swarm (but note that a swarm of bees is a closed group); in the case of fish it is called a school. Schooling occurs in

25 percent of fish species. A flock may be temporary, as with the migrating flocks of birds and the schooling of juveniles common among fish, or persistent, as with the obligatory schools of many marine teliosts.

Flocking, swarming, and schooling have several functions, which vary from species to species. For birds the two functions most often mentioned in the literature are foraging efficiency and predation protection: individuals that have located food can lead the other members of the flock to it, and the many eyes of a group can provide early warning of approaching predators. Collective defense can negate a menace that would be hazardous to isolated animals, and crowd behavior can cause a "confusion effect," making it difficult for a predator to fix on a single victim. Defense against predation appears to be one of the main functions of fish schooling also. Another function applying in at least some cases is synchronization: schooling can ensure that males are ready to fertilize the eggs when females are ready to spawn.

Focal animal sampling An observational sampling method in which a scientist follows the behavior of one individual in a group, and its interactions with other group members, exclusively for a certain time. At the end of that time another member of the group may be chosen as the focal animal. Concentrating on one animal at a time allows for a more detailed record of behavior sequences and interactions than directing attention haphazardly to the group as a whole.

Follicle rupture   See Ovulation

Follicle-stimulating hormone (FSH)   A gonadotropic hormone secreted by the anterior pituitary gland, which actuates growth of ovarian follicles and secretion of estrogen.

Follower   A designation sometimes given to the precocial young of many ungulates. From birth onward they show a drive to follow the mother (or a substitute moving object) and stay close to her. In this they contrast with lying-out young, which stay where their mother deposits them even when she moves away. Gnus, sheep, chamois, goats, and wildebeest have following young.

Following behavior In species that have parental care, a general term for the attempts of young to stay close to one or both parents. For example the fry of many species of fishes swim after both parents or the parent that does the brooding for several days (the length of time varies from species to species) after hatching (or leaving the mouth in mouth breeders). Following behavior is very pronounced in the young of many precocial species of birds and is usually referred to as the following response. Sometimes in the literature "following behavior" is used very broadly to include all behavior "with which a chick responds to the presence or absence or a parent, usually the mother, such as following, keeping close, giving greeting calls in the one case, crying and searching in the other" (Weidmann 1958). Following also occurs in many ungulates; see FOLLOWER.

Because the following behavior of young animals can be easily quantified and can be elicited in various ways with the use of stimulus models, it has been a popular subject of ethological study, especially in work on filial imprinting and the analysis of stimulus situations.

Following response See Following behavior

Following-response imprinting See Filial imprinting

Food begging Behavior patterns by which young animals induce their parents to feed them. Food begging presents the releasers required for eliciting and guiding the parental food-giving responses. Food-begging patterns include certain movements (wing fluttering in many young songbirds), body postures (begging postures), and vocalizations (begging calls), often in combination. In many species, adults during courtship perform elements of food begging (see COURTSHIP FEEDING), an example of change of function. In some of the more social species, such as honeybees and African wild dogs, food begging occurs among adult group members as a means of sharing food.

Zoo animals sometimes exhibit an artificial form of begging behavior: individually conditioned motor habits, often highly stereotyped.

Food hiding See Food storage

109

Food Imprinting The formation of a specific food preference, which is like imprinting in its rapidity and durability. It occurs mostly in highly specialized species, for example the larvae of plant parasitic insect species. Feeding preferences have been reported for some vertebrates (European polecats, domestic chicks, and various species of snakes and chelonians), in which experiments have shown that the preferences are established very quickly and very early, sometimes even with the first intake of food, in a manner that suggests a sensitive period of imprinting. However, such preferences may persist for only a limited time, so the appropriateness of the term "imprinting" is questionable.

Food provisioning Feeding of one animal by another. It occurs in the following forms: feeding of young by parents as part of parental care; feeding of females by males as part of courtship or pair-bonding behavior; and sharing of food by members of the same colony or hive, as in many species of social insects. Among some of the social insects division of labor provides for the feeding needs of individuals that are involved in activities such as defending the colony or tending the brood; in some species certain castes are incapable of getting food and are therefore entirely dependent on being fed by fellow members of the colony. Many insects transmit a pheromone with the food. Thus the dispensing of food can serve the additional function of disseminating chemical information (see TROPHALLAXIS).

Food is provided to members of other species in the exploiting of captive bugs by ants and the exploiting of hosts by brood parasites.

Food storage The laying by of food for use during times of insufficient supply. Birds (notably woodpeckers, corvids, titmice, and nuthatches), mammals (especially rodents), and some social insects store food. True storage of food requires two conditions: during some part of the year there must be an overabundance of available food, and the food must remain edible for the length of time it is hidden in the ground (as by squirrels and jays) or hoarded in a chamber (marmots, hamsters). Fruits, seeds, and nuts best meet these requirements. The various species of jays known for food storage have revealed a remarkable degree of cognitive capacity in remembering where they have deposited food items.

Food is also stored for short periods, especially by carnivores when they make a kill and are unable to consume it all immediately.

Form analysis   Close comparison of behavior patterns as a means of working out their possible evolutionary origins. Form analysis has been applied both within the behavior repertoire of a species and between the repertoires of closely related species to arrive at ideas about the course of ritualization in the evolution of displays from nonsignaling behavior. This conception of display as a "derived activity" has led to its interpretation as an elaboration of intention movements, ambivalent behavior, displacement activities, and redirection. The evidence from form analysis may be supported by the results of situation analysis and sequence analysis.

Form constancy   In ethology, the characterization of a movement that is always performed in the same manner. At one time it was considered as an indication that the behavior was innate (see FIXED ACTION PATTERN). Now it is recognized that acquired behavior patterns can be just as rigidly stereotyped as unlearned reflexes. The best-known cases of this are from caged animals, but there are many "natural" instances as well. For example, European songbirds that have been naturalized in New Zealand for more than a hundred years still sing like their European ancestors. In at least two of the species—blackbirds and chaffinches—it is known that a large part of the song repertoire has to be learned. Here form constancy has persisted over many generations, even though the behavior is not innate but transmitted through tradition. A high degree of form constancy is characteristic of behavior produced through ritualization.

Fossils   The preserved remains of past animals and plants or traces of their existence, such as footprints or burrowings, in rocks. Fossils are an important aid in the study of phylogenetic relationships but have not been very useful in the phylogenetic approach to behavior (see BEHAVIORAL PHYLOGENY).

Among the few "behavioral fossils" we have are trace fossils, such as petrified footprints, which may indicate the locomotion pattern of an extinct animal species; remains of constructions such as spiderwebs and termite nests; and preserved stomach contents and feces, which can lead to inferences about feeding habits and diet. Thus the prey catching of predatory marine snails (Naticidae or moon snails) has been traced back about 100 million years through study of the fossilized shells of their prey. In the course of this period these molluscs showed an early "progressive" development in their attack, but they have remained virtually unchanged in this regard for about 60 million years. In the earliest prey fossils the bore holes through which the predator inserts its proboscis and radula (rasping tongue) are distributed more or less randomly on the surface of the shell, with a slight preference for the aboral area (away from the opening). Later the bore holes are progressively concentrated on the adoral surface near the opening. Presumably this location is the most effective, perhaps because the predator can best elude the defensive efforts of the prey. This position was reached by the beginning of the Tertiary, and since then their prey catching has shown no further change.

Such evidence of behavioral evolution over geological time is available for only a few cases, and "behavioral paleontology," sometimes referred to as paleoethology, is a very meagerly developed branch of science.

Foster rearing   The raising of young by individuals other than the natural or biological parents. If this be done by individuals of a different species, it may be referred to as interspecific fostering, or cross-fostering, or cross-species rearing. When young animals are raised by people it is called hand-raising. As a rule only humanly contrived situations are referred to as foster rearing; hence brood parasitism is usually excluded, and if parental substitution occurs in other natural conditions, the more common term is "adoption."

Foster rearing serves primarily to preserve the lives of young that have lost their parents or have been lost. It is sometimes used to try to rehabilitate a species whose parental behavior has been disturbed by human management. In studies of the communicative significance of social stimuli, animals are foster reared to withhold experience of such stimulation during early life (see ISOLATION EXPERIMENT). Foster rearing has been very useful in studies of imprinting, especially sexual imprinting. Domestic animals are suited for this purpose because in the course of domestication, parental readiness tends to increase and so predisposes adults toward foster rearing.

One consequence of foster rearing, particularly hand-raising by humans, may be symptoms of deprivation in the young. Recently the transplantation of mammalian embryos has been successfully carried out in some cases, thus effecting an early start to the foster rearing process.

Fratricide Sibling slaughter; killing of nest mates. It is best known from birds of prey, such as hawks and owls, in which, because the eggs hatch successively at intervals of several days, there are considerable size differences between the siblings. Fratricide has also been described in fish and insects. Its function is not fully understood. Like infanticide, it probably contributes to adjusting brood size to food supply. Individual young are likely to be sacrificed when, because of adverse living conditions, a reduction in the number of young aids survival of the rest.

Fraser Darling effect Acceleration and synchronization of reproductive state in a breeding colony as a consequence of communal social stimulation. The effect is greater as group size or density increases. Darling (1938) first proposed this effect after studying the timing of breeding in different-sized colonies of herring gulls. An alternative explanation of localized breeding synchrony in colonial animals is that individuals in the same reproductive state attract one another to the same area. This is sometimes called the Orians effect after Orians (1961), who first suggested it (see COMMUNAL COURTSHIP).

Frequency modulation Variation of the wavelength of energy transmitted in wave form. In bioacoustics and animal behavior literature the term usually refers to sound, the frequency modulation of which is experienced as change of pitch. Melody is thus an effect of frequency modulation. Compare AMPLITUDE MODULATON.

Frustration Thwarting. Failure of goal-directed behavior to realize its objective, through obstruction or absence of the requisite conditions. In some psychological contexts the connotation includes the emotional state consequent on the thwarting of action.

Frustration-induced aggression Attack behavior consequent on frustration. For example, when a pigeon is switched from constant reinforcement to partial reinforcement in an operant situation, the bird turns aggressively to pecking at bystanders within range. The aggression is believed to be caused by aversive arousal arising from the situation and thus to be on a par with aversion-induced aggression, but it also bears some resemblance to redirection.

Function Adaptive significance; survival value. The question of the "ends served" by species characteristics, including behavior, is of great importance to biology as a whole and ethology in particular. However, there is some variation in the use of the term "function," which can give rise to misunderstanding. Generally in behavioral studies the function of a behavior pattern is taken to be the contribution it makes to survival and reproductive success (see FITNESS), that is, the specific consequences of its performance that in the course of phylogenetic history have led to its being preserved and shaped by natural selection in the repertoire of the species. This sense might be termed ultimate function, to distinguish it from the sense of "function" that refers to the immediate consequences of action, which might be termed the proximate function. For example, a proximate function of birdsong is territorial defense, which may mediate the ultimate function of adjusting population density to resource availability. The distinction is worth making because the methods used to elucidate proximate function often differ from those applicable to ultimate function, and the results of the one sometimes have little bearing on the other.

Yet another sense of "function" sometimes encountered in ethology derives from the use of the term in mathematics. When different measures vary together in a manner such that the value of one appears to be dependent on the value of another, the first is said to vary as a function of the second.

Functional system Functional cycle; behavior system. Von Uexküll (1957) coined the German term *Functionskreis* for the relationship between specific signs in the environment, their per-

114

ception through an animal's sense organs, and the (preformed) re-actions elicited in the animal.

In ethological literature functional system today usually means a set of behavior patterns serving the same or similar functions, for example, locomotion, foraging, courtship, parental care, or aggression.

GROOMING

G

Gamete Germ cell; reproductive cell. In most animal groups there are two kinds of gamete: motile semen cells (in males see SPERM), and nonmotile egg cells (ova) in females. At fertilization sperm and egg fuse to form the fertilized egg cell or zygote.

Game theory A part of applied mathematics dealing with situations in which the probabilities of "winning" and "losing" vary with the strategies chosen by the players. Ethologists have applied game theory to situations in which contending animals have alternative courses of action open to them, such as attacking, fleeing, and displaying in agonistic contexts. From estimations of the costs and benefits for each alternative combination (for example, the benefit of winning opposed to the cost of injury when both animals attack), the theory predicts the optimum proportions in which the strategies will occur. See also EVOLUTIONARY STABLE STRATEGY; SELECTION.

Gause's principle See Competitive exclusion principle

Gene The basic material unit of heredity, the means by which characteristics are transmitted from parents to offspring. The term presents a good example of how scientific concepts evolve along with the advances in understanding to which they contribute. A Danish geneticist, Wilhelm Ludwig Johannsen, coined the word "gene" in 1909 to refer to the factor called for by the facts of Mendelian inheritance. He explicitly rejected any morphological connotation. However, with the discovery of relationships between the transmission of hereditary traits and the distribution of chromosomes, and then between linkage of traits and chromosomal crossing-over, the gene came to be identified with a chromosomal locus. As molecular genetics unraveled the chemical structure of the deoxyribonucleic acid (DNA) strands of chromosomes, genes were identified with portions of such strands. As the biochemical nature of the gene has become clearer, it has become more difficult to pin down the concept in a single definition on which all can agree. However, drawing distinctions at the molecular level is not a concern to ethologists in most contexts in which they use the term. They can regard the gene as the minimum unit of replication in genetic transmission.

In this sense a gene is understood as existing in alternative forms or alleles, which may show a dominant-subordinate relationship when coupled in diploid organisms. Contrary to the earlier views of a one-to-one correspondence between genes and hereditary traits, it is now known that most traits are affected by multiple genes (pleiotropism) and that most genes (polygenes) affect more than one trait. Indeed the old "bean bag" conception of genetics has given way to the conception of genes integrated in a gene complex, as natural selection either weeds out disruptive changes, called mutations, or leads to ways for them to be accommodated. Mutations may involve either single genes or sections of chromosomes.

Gene pool The totality of genes in a population of interbreeding organisms; all the hereditary factors available for segregation and assortment within a reproductive assemblage. Changes in the composition or proportional representation of genes in a gene pool from one generation to another, which may result from selection, genetic drift, or mutation pressure, constitute evolution. Compare HARDY-WEINBERG LAW.

General adaptation syndrome   See Stress

Generalization   See Stimulus generalization

Genetic drift   A change in gene frequency in a population because of chance sampling independent of selection. In a very small population the likelihood that certain genes will be lost and others fixed by chance increases through the random assortment process of reproduction. Evolutionary biologists have had differing views about the importance of genetic drift as a force in evolution.

Genetics   The science of heredity; study of the transmission of hereditary characteristics from generation to generation. It consists of several parts. Classical or Mendelian genetics uses selective breeding to determine the patterns of distribution of hereditary characteristics between parent and descendant generations. Molecular genetics deals with such questions as the biochemical nature of the gene, the genetic code and the means by which its information is drawn on and used at the cellular level, chromosomal and gene mutations, and the replication mechanisms at work in cell division. Population genetics is concerned with how gene frequencies change or become stabilized within populations, and hence with selection, mutation rates, and genetic drift. Developmental genetics has to do with how genetic information enters into ontogenetic processes from formation of the zygote onward. Behavioral genetics studies the inheritance of behavior patterns and dispositions.

Genital display   Use of the genitals for signal purposes. It is best known from primates, in which it takes two main forms. Females, and subordinate males in some instances, turn the hindquarters toward the other animal as a form of soliciting behavior or appeasement behavior. This is generally referred to as presentation or genital presentation. The display is frequently made conspicuous by the morphological support of swelling and coloring of the bare skin surrounding the genital and anal region. The other genital display is of the penis, which is usually presented in erection and used as a threat signal. In some cases, as in marmosets (*Callithrix jacchus*), the males combine penile presentation with urination and marking behavior. In the vervet monkey (*Cercopithecus aethiops*) the penis is bright red and the scrotum bright blue—colors that add vividly to the genital display.

Genome Genetic endowment; the totality of an individual's hereditary determinants. The genome thus consists of the genetically encoded information governing formation of body structures, physiological processes, and the mechanisms controlling certain kinds of behavior or behavioral components (see INNATE).

Genotype The genetic constitution underlying a particular trait or group of traits. It is coupled with the term "phenotype," which is the realized body form or physiological process to which the genotype gives rise in interaction with the environmental conditions present during development (see EPIGENESIS). Because of genetic dominance, different genotypes can give rise to the same phenotypic character, and because of environmental variation the same genotype can give rise to different phenotypic characters.

Geotaxis See Taxis

Germ cell See Gamete

Gestalt perception The capacity of many animals to apprehend specific stimulus combinations, not only as the totality of their component characteristics (see STIMULUS SUMMATION), but also the particular relational structure among the components (configurational stimulus). Examples of such transsummation, in which "the whole is more than the sum of its parts," occur in acoustic as well as optical perception. Experiments have shown that songbirds whose songs have long and varied note sequences base their reaction to an artificial song on the whole acoustic pattern rather than on the component characteristics taken separately when presented with a playback of versions of the song. Also many visual patterns are effective only when all the components are present and arranged in a specific configuration with respect to one another. This holds for the recognition of star patterns by which nocturnal migrant birds determine their orientation during migration.

Gesture Expressive movement or posture of the body involving the torso, extremities, and appendages (for example, the tail). Compare FACIAL EXPRESSION, which refers to expressive appearance of the face.

Gonad Reproductive organ. The male gonads are called testes; the female gonads, ovaries. They have two main functions: production of gametes (sperm and egg cells) and production of sex hormones.

Gonadectomy Castration. Removal of the gonads. If the operation is done on a male it may be called orchidectomy, orchiectomy, or emasculation; on a female, ovariectomy. In behavioral research gonadectomy is an important method in studying the effects of sex hormones. As a result of recent advances in biochemical technology, gonadectomy can now be effected with antihormones and other sex hormone inhibitors or antagonists, which avoids having to use surgery, with its associated drawbacks, and may also offer the convenience of reversibility of effect.

The concept of castration has recently undergone an interesting extension in ethology. In many socially living mammals, for example viverrids, only the highest-ranking pair in a group reproduces at any one time, even though other members may by physiologically ready to breed and in fact will breed if the highest-ranking pair is removed. Also in primates, suppression of breeding capability in subordinate males by the mere presence of higher-ranking rivals has been observed. Such cases have been described as instances of psychological castration (see SOCIAL INHIBITION).

Gonadotropic hormone Gonadotropin; gonadotrophic hormone; gonadotrophin. In vertebrates, hormones produced by the anterior pituitary gland (adenohypophysis) that influence the activity of the gonads and thereby the secretion of sex hormones. They comprise follicle-stimulating hormone (FSH), luteinizing hormone (LH), and interstitial-cell-stimulating hormone (ICSH).

Graded potential See Graded response

Graded response A response that varies in proportion to the strength of the stimulus evoking it. For example, the generator potential (graded potential) produced in a receptor is typically a direct function of the intensity of its adequate stimulus (see RECEPTOR), in contrast to the action potential produced by a nerve cell, which is an all-or-nothing response.

Graded signal   A display that varies in intensity of form as a function of underlying motivational state. A classic instance is the variation in body markings that correlates with the level of sexual arousal in the guppy. Compare TYPICAL INTENSITY.

Grasp reflex   See Clasp reflex

Grooming   Care of the hair covering (pelage, fur, coat) and skin by mammals; the mammalian equivalent of preening in birds. Grooming includes scratching, rubbing against objects, licking, nibbling, rolling in dust, and washing. In the primate literature it sometimes means social grooming. To avoid the possibility of misunderstanding, the preferred terms are the more cumbersome "allogrooming" for an animal's manipulation of the fur or skin of another animal, and "autogrooming" for an animal's attention to its own body. See also SOCIAL GROOMING.

Group cohesion   Continuous association of a number of adult animals, often with their offspring, over a period of time. It is especially pronounced in animal groups showing closed sociality (see GROUP FORMATION). Many social animals have developed special means of achieving group cohesion such as certain vocalizations, group odors, and appeasement behavior, that strengthen attachment among group members and suppress disruptive aggressiveness. It is possible that group cohesion depends upon motivation special to it and independent of other drives. See BONDING BEHAVIOR; BONDING DRIVE.

Group courtship   See Communal courtship

Group effect   All the benefits accruing from group living, such as antipredator vigilance (see WARNING BEHAVIOR), collective defense, cooperative foraging, and synchronization (see FRASER DARLING EFFECT).

Group formation   In many species, over and above associations in pairs and families, larger numbers of individuals gather in temporary or long-lasting groups. Such groups vary widely in the types of kin relationships of their members and in their age, sex, and even species composition. Most groups consist of a single species, but some, referred to as mixed-species groups, are composed of two or more species. In the simplest case groups form "by accident"—as a consequence of a number of individuals independently seeking the same kind of place and converging. Such fortuitous gatherings are described as pseudosocial groups or aggregations. More advanced in a social sense are groups based on social attraction; that is, the members have sought one another out or aimed toward one another. For the more highly organized of such groups ethologists have adopted the term "society." Social groups can be differentiated into open societies, which tolerate newcomers and allow interchange with other groups, and closed groups, in which the members discriminate against outsiders, excluding them and hence the possibility of interchange. Open sociality occurs predominantly in species that migrate in groups for long distances. In mammals such groups are usually called herds, sometimes flocks; massed insects are referred to as swarms; birds move around together as flocks, and fish as schools. Closed societies occur in two forms: those in which the members recognize group-specific characteristics but cannot discriminate between individuals within the group and those in which the members react to one another on the basis of individual recognition (see ANONYMOUS GROUP; INDIVIDUALIZED GROUP). Anonymous groups include insect societies and the colonies of many rodent species. Individualized groups occur in many primates and some carnivores (wolves, African wild dogs, lions). Usually the group members have a rank order. Delineation of group membership can present difficulties for field work when, as often happens with primates, several groups converge at one locality, such as a watering hole or sleeping place. Observers then use the proximity of individuals to one another, and the frequency and kinds of interactions between them, as indices of group affinities, a group being "a number of animals that remain together in or separate from a larger unit and mostly interact with each other" (Kummer 1971).

Group mating   See Communal courtship

Group odor   Family smell; colony odor. A group-specific smell established by reciprocal scent marking or odor exchange in a group
of communally living animals. It is especially common in mammals
and social insects. Such marking provides a "membership certifi-
cate"—a means by which the members of a group can tell one
another from strangers, even though they do not recognize one
another individually. Group odors are exemplified by some rodents,
the marsupial flying phalanger, and honeybees. See HIVE ODOR;
MARKING BEHAVIOR; PHEROMONE).

Group selection   Selection among groups of interbreeding organisms within a species, as opposed to selection among individuals.
Wynne-Edwards (1962) argued at length that group selection has
been of the first importance in the evolution of social behavior,
especially behavior that regulates population size. Part of his theory
was that the intensity of performance of certain displays, which he
called epideictic displays, is somehow correlated with food supply
and the expected demand for food in the coming breeding season,
and so determines how much breeding there can be. In anticipation
of too many mouths for too little food, limitations on population
growth are set by such measures as curtailment of breeding by some
individuals, socially induced mortality, and delayed sexual maturation. Because the genes underlying such strategies benefit the group
as a whole, but not the individual in competition with other members
of the group, the social evolution is governed by competition and
selection between groups, not between individuals.

Since the publication of this theory, evolutionary biologists have
debated the issue of group versus individual selection; one outcome
has been the emergence of sociobiology. It now appears that group
selection in the form envisaged by Wynne-Edwards requires very
special conditions, which can rarely be realized. Much of the evidence he used for it has been reinterpreted as consequent on kin
selection.

Group territory   Communal territory. A territory held in common
for a long period by a number of conspecifics, for example a harem,
a multimale group, or a long-lasting cohesive family group.

Guarding behavior   Mate guarding; "wife watching." Guarding behavior was first observed in the efforts of harem-forming male primates and herding male ungulates to maintain continuous control over and exclusive access to the females in their possession. Especially striking was the guarding behavior of hamadryas baboons: females that stray too far from the male are threatened by him and punished with bites if they do not come immediately to heel. More recent observers have reported various forms of male vigilance toward females in several other taxonomic groups, including mere attendance on the females, maintenance of continual bodily contact, and active warding off of other males. In some species of dung flies the male stays with the female after mating and prevents the copulation attempts of other males until the female has laid her eggs. Similar behavior has been seen in many dragonflies and damselflies. In numerous frog and toad species the males stay clasped to their females all day long, and in many birds (swallows, magpies) the males stay much closer to the females during the time of egg laying than at other stages of the reproductive cycle. According to these observations guarding behavior is most intense at times when mating has the greatest chance of effecting fertilization. Its biological significance therefore apparently consists in prevention of copulation with other males, hence avoidance of sperm competition, hence securing of reproduction of the guardian male's genes.

The terms "guarding" and "vigilance" are sometimes applied to watching over conspecifics other than the female, as, for example, when a father or other group member picks up an infant and holds it in a threatening situation.

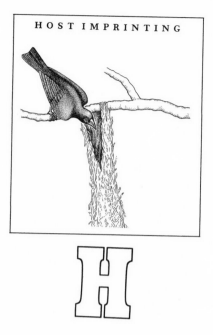

HOST IMPRINTING

H

Habit An individually acquired stimulus-response connection; a behavior pattern consequent on individual experience. The term often carries the connotation of recurrent action done without thought or attention, at least for human behavior. In psychology it has much the same sense as "conditioned reflex." However in some contexts the term is used more or less in the same sense as "behavior" or "behavior pattern," as when people speak of the habits of a certain species.

Habitat The kind of place in which a particular organism is typically found; the kind of environment (including climate, vegetation, and other factors) to which the organism is adapted. In German, *habitat,* in its strict sense, pertains to a single species, in contrast to *biotope,* the environment occupied by a natural community, which pertains to numerous animals and plants. However, in English the distinction is not so sharply drawn.

Habitat imprinting Environmental imprinting. A preference for a specific type of habitat as a result of early experience. In many species the breeding habitat is chosen on the basis of similarity to the habitat in which the animals were reared. Exchange experiments, for instance with various bird species, have shown that such a preference depends upon learning during a sensitive period in early life, the influence of which can persist throughout life. This kind of learning thus meets two of the main criteria of imprinting as classically conceived.

Habit preening   A form of ritualized preening movement common in the courtship behavior of ducks. It derives from the movement in which a duck turns its head over its shoulder to tend the feathers on the inside of a folded wing. In some species its conspicuousness as a signal is enhanced by morphological support. In the mandarin duck, for example, the movement tilts up an enlarged and brightly colored feather, and in the garganey the bill points to a brightly colored patch (see SPECULUM) on the outside of the wing. Habit preening is an example of ritualized displacement activity.

Habit strength   In behavioristic learning theory, a technical term defined as the number of times a stimulus-response pairing has been reinforced.

Habituation   Stimulus-specific waning of response; learning not to respond to something on finding that nothing significant is contingent upon its occurrence. More roughly, getting used to something to the extent of ignoring it, getting tired of it, becoming inured to it. The stimulus specificity of habituation is demonstrated when a new stimulus, replacing the one to which response has become refractory, immediately reinstates the response at approximately full strength. This shows that neither muscular fatigue nor sensory adaptation caused the decline in response. If sensory adaptation were responsible, response to the stimulus would return in a relatively short time, but the effect of habituation usually persists considerably longer. This distinction appears inapplicable to some of the lower animals, such as annelid worms, whose relatively poor discriminative abilities would make long-lasting habituation maladaptive: an animal that cannot tell the shadow of a waving weed from that cast by a roving predator is going to be in trouble if habituation to the first lasts into an occasion of the second. Indeed, experiments involving repetitive presentation of stimulus models have shown that waning of responsiveness may be a complex process consisting of various changes differing in kind, duration, and specificity.

Writers are not always consistent in their use of the term "habituation" and hold differing views about its meaning. Some writers regard habituation as a simple form of learning, but others object that no true learning occurs because no new association is formed, only an old one weakened or removed. In reply it is argued that habituation enters new information—the lack of significance of the stimulus—into memory and qualifies as learning on this account. Habituation can thus be viewed as the mirror image of classical conditioning, since it turns an effective stimulus into a neutral stimulus instead of the reverse.

The term is also used, although rarely in ethology, in a sense close to that of "addiction": people or animals are said to have become habituated to a drug, meaning that they have become dependent on it.

Hamilton's rule If an allele that produces altruistic behavior causes an animal carrying it to act toward relatives in such a manner that the product of the benefit to these relatives and their relatedness to the animal is greater than the cost to the animal (with benefit and cost computed in terms of reproductive success), the allele will spread and become established in that population (Hamilton, 1975). The condition for kin selecton can thus be expressed as $rb - c > 0$, or $b/c > 1/r$, where r stands for relatedness, b for benefit, and c for cost. In practice, relatedness can be determined from genealogical data, electrophoretic analyses, and some other kinds of evidence; the simplest measure of benefit and cost is the number of offspring. The rule is implicit in the concept of inclusive fitness.

Handedness A preference for the use of the right or the left hand or foot. Examples of handedness are known from rats, cats, and various primates, and from birds, such as parrots, that use their feet to convey food to the bill. In primates handedness is more strongly developed in ground-living species, such as baboons and macaques, than in arboreal species, such as vervet monkeys, probably because an extreme right-left specialization would be disadvantageous for locomotion among branches. We do not know why handedness

127

evolved or what its biological significance might be, although functional specialization along the length of the body (for example, in kangaroos, small forelimbs for manipulating and large hind limbs for locomotion) might give a clue. In humans handedness is associated with lateralization of function in the forebrain.

Hand-raising   See Foster rearing

Haplodiploidy   In hymenopterous insects (ants, bees and wasps) an individual's sex is determined by whether it develops from a fertilized or an unfertilized ovum. Fertilization supplies two sets of chromosomes, and the resulting diploid insect is female. The unfertilized egg has only a single set of chromosomes, and the haploid insect arising from it is male. An interesting aspect of this unusual mode of sex determination is that it can reconcile the existence of sterile worker bees with the theory of natural selection. The workers in a hive are female and therefore diploid; they are also sisters, which means that each one carries the genes of her (haploid) father and also shares roughly half of the set of genes she got from their mother. A worker bee thus has 75 percent of her genes in common with her sisters, and would have only 50 percent in common with a daughter, so she increases her inclusive fitness more by promoting her fertile sister's reproduction than by being reproductive herself.

Hardware   See Program

Hardy-Weinberg law   If in a population the frequency of one allele, a, is p, and the frequency of another, a', is q, then on the average p^2 individuals will be aa, $2pq$ will be aa' and q^2 will be $a'a'$. This will be the same from generation to generation unless these frequencies are affected by emigration, selection, mutation, nonrandom mating, or sampling errors. In other words gene frequencies in populations of randomly breeding individuals will remain the same unless evolutionary forces act to change them.

Harem   A stable and lasting association of one male with several females. Harem formation is a type of polygamy. Harems are found in mammals (zebras, some antelopes, and numerous primates, such as patas monkeys, geladas, hamadryas baboons, and langurs), many lizards and fishes (African and South American cichlids).

Some temporary associations of a male and several females are described as harems, for example, in the pinnipeds (seals, walruses). However, these animals stay together only as long as it takes for mating to occur. Also the composition of the female contingent may change continually, and the male's place be taken over by another male. The term "harem" is not really appropriate for such mating systems, which are more aptly described as instances of promiscuity.

Head scratching Early in the history of ethology, the manner in which birds use their feet to scratch their heads was described and frequently discussed. Two patterns of such scratching occur: "around the front" scratching, in which the foot is carried up in front of the breast to the head; and "over the shoulder" scratching, in which the wing on the same side as the scratching foot is lowered, and the foot passed over the shoulder joint to reach the back of the head. This awkward-looking movement sequence, which often causes young birds to lose their balance, makes clear sense only for animals with the body proportions typical of tetrapods, in which the forelimbs are forwardly placed and the hind feet can make contact with the head only by going over the shoulder. Nevertheless, the movement is remarkably pervasive, even in animals with the motor capability to perform the action in the more direct way.

The suggestion to account for this, which was adopted in much ethological writing, was that over-the-shoulder scratching is a heritage from reptilian ancestry (see RELICT). Two facts argue against this interpretation. First, the distribution of the two forms of scratching is more or less random between primitive and advanced bird groups, giving no indication of which form is the phylogenetically older. In some cases different scratching patterns occur within related groups and even between closely related species. Second, in some species the young birds differ from adults in their scratching patterns, and in these cases the young scratch around the front. According to argument based on the notion of recapitulation, the direct method should thus be the older one. The question of over-the-shoulder scratching in birds has yet to be settled.

Heat Typically used of female carnivores, such as cats and dogs, to refer to the periodic state of sexual receptivity and arousal. It is therefore synonymous with estrus as applied to these animals. It is also more loosely used for sexual arousal in mammals in general.

Helper A term used mainly in the ornithological literature for individuals that assist parents in the raising of young. They are often also called brood helpers, or helpers at the nest. As a rule helpers are relatives of the parent pair, such as offspring from a previous brood or the previous season, but in some species the helpers are not close kin. Among birds this behavior is widely distributed: so far it has been described for more than 150 species in various orders. It has apparently evolved numerous times (see CONVERGENCE).

On the question of the biological significance of helping, very different conjectures have been made. At first it appears as a classic form of altruism, yet, as with warning behavior, it is possible that the helpers are acting entirely from self-interest. The helper may benefit, at least incidentally, from the experience of carrying out parental care before it breeds for the first time. Also procuring a home in the parental territory may be a boon in a number of ways, including the advantage it may give to a male if opportunity arises to take over the territory or split off part of it for himself. The variations in number and kinship relations of helpers, the extent and duration of their brood-care activities, and other features make it unlikely that a single functional explanation could cover all instances.

However, it is quite generally supposed that the direct individual benefit to a helper must be greater the less closely related it is to the parents and young it helps, because the less the degree of kinship the less the helper benefits indirectly from the addition to its inclusive fitness of promoting replication of its own kinds of genes; nonrelatives receive no indirect fitness benefit (see KIN SELECTION). There is evidence in a few cases that the helpers do not really help, but, on the contrary, may harm those they tend.

Behavior similar to that of the brood helpers among birds is found in other animal groups, although the term "helper" may be less appropriate in such cases. Examples include many primates, in which childless females (see ALLOMOTHER), older sisters, and other group members temporarily take care of infants; and dolphins, in which group members raise newborn infants to the water surface to breathe. Helping is also known in fish, such as the brood-caring African cichlids, and in amphibians. In the social insects brood care may be largely or exclusively carried out by individuals other than the biological parents, although closely related to them (see CASTE).

Herd   A form of open social group (see GROUP FORMATION). The term, usually reserved for ungulates, corresponds roughly to "flock." Similar but usually smaller groups of carnivores are referred to as packs.

Herd instinct   See Bonding drive

Heredity   See Genetics

Heritability   The proportion of the variance of a trait in a population that can be attributed to genetic inheritance as opposed to environmental influence. Thus eye color in people has a heritability score close to one because it is almost totally determined by heredity; a person's spoken language has heritability close to zero because it is a consequence of experience.

Heterogeneous summation   See Stimulus summation

Hibernation   A state of torpor that many animals enter into during winter months. Metabolism is markedly reduced, and body temperatures may drop to near zero degrees centigrade in some cases. Physiological adaptations include the lowering of the freezing point of body fluids. Behavioral adaptations include selection and preparation of a place in which to hibernate, such as a den or burrow that provides some insulation and may protect against dehydration. Hibernation and the less extreme forms of dormancy, such as that found in bears, enable animals to survive through severe winter conditions and are thus alternatives to migration. It occurs in a variety of insects, terrestrial molluscs, amphibians, reptiles, birds, and mammals.

Where the climate is periodically intensely hot, some animals enter a similar form of dormancy, called estivation, to tide them over.

Hierarchy   Aside from its use in taxonomy, the notion of hierarchical organization has two main applications in ethology:

First, applied to social organization, hierarchy is synonymous with rank order and involves the concept of social dominance. Individuals in a group yield to others when in contention for something, such as food or a mate, according to a more or less linear order, with the top individual dominating all the others and the bottom individual being dominated by all the others.

Second, applied to motivation, hierarchy means that control of behavior is a system of tiered or nested units distributing causal or functional management in a stepwise manner, with the most general at the top and the most specific at the bottom. The best-known conception of this is Tinbergen's (1951) hierarchical theory of instinct, but there have been numerous others, including some neurophysiological varieties. It has been argued that some theory of hierarchy will be most likely to synthesize ethological understanding of behavioral control (Dawkins 1976).

Yet another way in which ethologists have employed hierarchical ordering is in the description of behavior, classified in terms of a set of inclusive categories: motor patterns grouped into classes, which are grouped into higher-order classes, which are gathered into still higher-order classes, and so on. The best-known example comes from Tinbergen's breakdown of the reproductive behavior of the three-spined stickleback into functional categories. However, confusion between such functional classification hierarchies and hierarchical conceptions of control often causes muddled thinking about motivation, as the history of ethology exemplifies.

Hitchhiking The use of animals of other species as means of transport. The technical term coined for this is phoresis, or phoretic behavior, but it can hardly be said to have achieved currency. Hitchhiking is most common among very small animals living on substrates, that are too far apart for their locomotor powers to manage. Thus many coprophagous mites and dung beetles get from one dung pile to another by hanging onto the fur or feathers of larger creatures that travel the route. Carrion-eating threadworms (nematodes) ensconce themselves under the wings of carrion beetles and so get taken to the feasts of their carriers. Such exploitation qualifies as a form of commensalism.

Hive odor The scent unique to each beehive. It enables the members of the hive to recognize one another and to detect foreign intruders. It is therefore similar to the group odor of some mammalian associations. However, in contrast to the group scent of mammals, hive odor is not a mixture of odorous body secretions, but a mixture of flower scents from the blooms upon which the bees have foraged (see PHEROMONE).

Hoarding See Food storage

Home range The area that an individual, pair, or group regularly occupies or repairs to, but which, in contrast to a territory, need not be defended against entry by other members of the species. Spatial separation is effected by the animals' simply keeping out of each other's way. Commonly, a home range includes a neutral zone between two or more territories, or a resource area impossible or energetically uneconomical for one animal or group to defend alone, such as a watering hole or occasionally a resting place or grazing area. The term is used mainly of mammals.

Homing The ability of some animals to find their way back to a specific locality, such as the birthplace or the breeding site of a previous season. It requires a highly developed orientation capability (see NAVIGATION). Especially impressive are the homing of migratory birds and many migratory fish, such as salmon. The homing capacity of pigeons has been the subject of intense investigation, and hence the source of much of what is understood about how birds find their way on long journeys.

Homology Correspondence with respect to relationship or position in a set or pattern. For example, the proximal bones of the foreleg of a horse, the wing of a bat, and the arm of a human are homologous, since they occupy the same position in the skeletal plan that is common to all tetrapod vertebrates; hence each of these bones is called the humerus. This criterion of homology in comparative anatomy was formalized as the "principle of connection" by Geoffrey St. Hilaire in the nineteenth century and elaborated in English by Robert Owen, who drew a firm distinction between homology and analogy, or similarity of form associated with similarity of function.

Homologies provided evidence for and were explained by the theory of evolution, which has since appropriated the connotation of the term; "homology" is now generally defined as similarity due to common ancestry. This can cause confusion because some features said to be homologous are not at all similar (for example, the amphibian quadrate and the mammalian incus) and because homology is sometimes used as evidence of common evolutionary origin.

This problem of definition is even more difficult with regard to behavioral homology. While homologies of structure are, in many cases, relatively easy to recognize from fossil intermediate forms, behavioral homologies can be sought only in living species, and no one criterion suffices to cover all cases. Several criteria have been used in plausible arguments for judgments of homology in behavior patterns, including the following: similarity of form, meaning that the finer the detail of resemblance the stronger the evidence (criterion of special quality); correspondence of position in a sequential pattern, such as an action chain (criterion of position, similar to the principle of connection in comparative anatomy); occurrence of graded intermediate or transitional forms in closely related species or in stages passed through in ontogeny (criterion of connection by intermediate forms); and independent evidence of common evolutionary derivation (criterion of common ancestry). The more criteria are met, and the more closely the details correspond, the greater confidence with which a judgment of behavioral homology can be made. For ethologists applying the criterion of common ancestry, the methods of cladistic taxonomy (see CLADISM) can be brought to bear on the distribution of a behavior pattern among related species. However, this will be of no help when phylogenetic relationship is at issue, or when the problem is to trace the course of ritualization from evidence of homological correspondence.

The term "homology" is sometimes applied to correspondences within a body structure or between cellular constituents: in segmented animals the parts of segments such as appendages in arthropods or segmental nerves in vertebrates are said to be homologous, and the relationships may be referred to as serial homologies. The chromosomes of a pair, and corresponding chromosomes in haploid sets in general, are referred to as homologous chromosomes.

Homosexuality Sexual behavior between individuals of the same gender (see PSEUDOCOPULATION). In natural conditions homosexuality has been observed in all-male groups of mammals with harem mating systems. This possibly serves to maintain a continuous readiness for copulation and so ensures an immediate reproductive capability as soon as an opportunity arises to take over a harem. Homosexuality has been induced experimentally in mallard ducks and zebra finches by keeping young males together with same-sex conspecifics during the sensitive period of their sexual imprinting. This brings about erroneous imprinting to the animal's own sex and so results in homosexual pairing. Occasionally homosexual behavior is observed between normally matured individuals when, as can happen in zoos, no opposite-sex partner is available.

Another case that has been described as homosexual is that of pairs of female birds incubating a clutch and brooding chicks like normal heterosexual pairs, the best known being "lesbian" gulls. However, in this case the homosexuality applies not to sexual behavior but to the parental arrangement only, which probably arises because there are insufficient males for each breeding female to pair with one, and it is impossible for a female to incubate eggs and brood chicks successfully on her own.

Hormonal manipulation   Hormonal treatment. Interfering with an animal's endocrine state by administering or removing hormones. They can be administered in various ways: orally, by putting the chemical in drinking water or food; by intravenous or intramuscular injection; or by implanting under the skin or in the brain. This latter method has the advantage of providing a means of diffusing a specific hormone continuously over an extended period. On the debit side, it cannot produce fluctuations in hormone secretion, which possibly play a role in the regulation of a physiological process or behavioral system. A hormone can be removed surgically by cutting out or otherwise destroying the gland producing it (see GONADEC-TOMY) or chemically by administering substances that block its synthesis in some way or prevent its action by occupying receptor sites in the target tissues.

Hormone   Organic substance produced by internal secretory glands (ductless glands) that is passed into the bloodstream and so transported to other parts of the body for various purposes such as regulation of metabolic and growth processes. The behavior of animals, especially behavior having to do with reproduction, is also influenced by hormones in numerous ways (see SEX HORMONE). A second group of hormonelike substances, the pheromones, are produced in the body but discharged into the environment, where they can affect the behavior or physiological state of conspecifics. See PHEROMONE.

Host imprinting This term is frequently used to refer to the relatively strong fixation of brood parasites on their host species. Thus it is known of the European cuckoo that each female always lays her eggs in the nests of the species that raised her (unless none is available), and the same has been reported for several African cuckoos. In the African widowbirds (Viduinae), a brood-parasitic form of weaver finch, each species or subspecies parasitizes a particular species or subspecies of estrildine finch. Here the specialization goes so far that the nestlings of the brood parasites conform to the young of their hosts in appearance (for example, in their mouth markings, which form part of the releaser for regurgitation feeding) and behavior (in their begging calls), and the widowbird males pick up the entire vocal repertoire of the host males and use it as their own. Although such features as the mouth markings imply a degree of genetic adaptation, the host fixation appears not to be innate but to be acquired while the young brood parasite is being raised by its host. The early learning phase and the manifest stability of the preference consequent on it conform to the two main characteristics of imprinting, so the term "host imprinting" appears to be appropriate. However, no experiment has tested whether the fixation is consistent with other connotations of the term.

Huddling Contact behavior. Close packing together of conspecifics, as in the roosting flocks of some birds (wrens, swifts, penguins). In many cases the animals position themselves so as to have as much body surface as possible in contract with that of a neighbor. Huddling serves to conserve heat, and in some species it appears to have acquired the additional or secondary social function of contributing to pair or group bonding.

Human ethology A relatively new branch of ethology concerned with investigation of the biological side of human behavior. In contrast to psychology, human ethology views behavior "from the outside," that is, through observation rather than through discussion and questioning. Furthermore, it concentrates on humanity rather than the individual, on general, pervasive behavioral characteristics comparable to the species characteristics of animals. It applies two of the main research methods of ethology: comparison of different species and populations and study of behavioral development. Comparative study can include observations of nonhuman primates, but as a rule it is confined to comparison of different human cultures. Cultures that have remained relatively primitive or so geographically isolated as to have had virtually no communication with the civilized world have been especially valuable to this comparative enterprise. The ontogenetic study concentrates on observation of newborn infants, who have had little or no opportunity for learning and on children who, because of a sensory deficit such as blindness or deafness, have persisting limitations on what they can learn.

Both approaches have revealed a number of widely distributed, largely experience-independent human behavorial characteristics, such as the laughing of children born blind and deaf, for which a predominantly genetic basis can be argued. Human ethology overlaps with some areas of anthropology.

Hybrid   Offspring resulting from cross-breeding between members of two different strains or species. Study of the behavior of hybrids plays an important part in behavioral genetics. As a rule interspecies hybrids are infertile and of low viability as a consequence of incompatibility between the genetic factors of the two parents. Under natural circumstances such hybridization is usually prevented in some way (see REPRODUCTIVE ISOLATION). The crossing of different mutant strains of fruitflies and the use of inbred strains of mice have contributed considerably to advances in genetics, including the genetics of behavior.

Hypersexuality   Hypertrophy of sexual activity. It is widely distributed in domestic animals but can also occur transiently in wild animals subjected to prolonged social isolation. Certain brain lesions, such as damage to the amygdala in cats, can cause hypersexuality.

Hypertrophy   In ethology this term is sometimes used of excessively frequent occurrence of a behavior pattern. In this sense it applies most often to domestic animals, particularly to their hypersexuality. It is the opposite of a behavioral deficit (see ABNORMAL BEHAVIOR).

Outside ethology, "hypertrophy" usually refers to abnormal growth of part of the body, and this sense occasionally applies in ethology as well as when overdevelopment of an endocrine gland has behavioral consequences.

Hypnosis A state of tonic immobility induced by restraint or the close proximity of a predator. Several theories have been advanced about the physiological basis and biological function of hypnosis in animals, but clear understanding of the phenomenon has so far proved elusive. There is some support for the view that it serves as a last-ditch possibility of escape for an animal in the grip of a predator (see THANATOSIS).

Hypophysis See Pituitary

Hypothalamus The ventral part of the anterior division of the brain stem (forebrain, diencephalon) in vertebrates. It contains areas involved in the regulation of such functions as eating, drinking, and reproduction. For example, neurosecretory cells in the hypothalamus regulate the output of hormones from the pituitary gland and thus affect the gonadal, thyroid, and adrenocortical functions, as well as exerting more direct effects through production of prolactin, oxytocin, and other hormonal substances.

Hypothetical construct See Intervening variable

I

Imitation Mimicking; observational learning. An animal's copying of a movement, vocalization, or modus operandi of another individual, either of its own or another species, and incorporating it into its own action pattern, vocal repertoire, or practical know-how. Imitation is one of the many manifestations of learning. It is especially well represented in the higher mammals, most notably in primates, and also in the vocal learning of many bird species, such as parrots and songbirds. Imitation of the vocalizations of other species is sometimes referred to as vocal mimicry or mocking (hence mockingbird). A process that is often taken for imitation, but which should be distinguished from it, is social facilitation. A completely different category of imitation is the kind of mimicry in which a species acquires features of the appearance or behavior of another "model" species in the course of phyletic evolution.

Implantation This term has two applications in physiological study. First, it can refer to the embedding of an embryo in the wall of the uterus prior to fetal development in mammals. Second, it can refer to the placing of hormone deposits, usually in capsule form, under the skin or elsewhere in the bodies of animals in hormonal manipulation experiments.

Imprinting   In ethology, a comparatively rapid kind of learning that occurs usually in young animals. Imprinting differs in two respects from other kinds of learning: it is restricted to a distinct sensitive period, and it is very stable, sometimes irreversible, in effect. The two classical examples of imprinting are the following response of precocial chicks and the sexual imprinting found in various species.

More recent study, especially in ecological contexts, has shown that these two characteristics can be found in some forms of learning not originally considered to be imprinting. Consequently these are now frequently referred to as cases of imprinting. Examples are the establishment of a persisting preference for a specific habitat type (see HABITAT IMPRINTING) and, in migratory species, fixation on a specific geographical locality. Many animals acquire early, rapidly formed food preferences (see FOOD IMPRINTING), and the rapid early song learning of many birds has been called song imprinting. Since the defining characteristics are less pronounced in many of these cases, a number of writers refer to them as merely imprinting-like or similar to imprinting. Moreover, the boundaries between imprinting and other forms of learning are no longer as distinct as they were once considered to be.

The imprinting idea has recently been applied to developmental processes in the central nervous system, which are like imprinting in behavioral contexts in that they occur very early, are very persistent, and are more or less impervious to later reversal. The terms used are "neural imprinting," and, where the determining factor is a hormone, "hormonal imprinting." Such processes in mammals play a role in determining whether an individual performs as a male or as a female in sexual interactions.

An instance of imprinting-like learning in adult animals occurs in sheep and goats: the ewes and she-goats get to know their lambs and kids individually by smell shortly after birth and on this basis will reject all others thereafter.

Incentive   In discussions of learning theory, the meaning of this term varies from one authority to another. In some contexts it refers to the arousing effect of a stimulus through its association with reinforcement (incentive-motivational effects); in other contexts it refers to arousal produced by withholding reinforcement or preventing attainment of a goal; in yet other contexts it is more or less synonymous with "appetite."

Incest avoidance A general term for mechanisms that obstruct mating between closely related individuals and so prevent the negative consequences of inbreeding (see ASSORTATIVE MATING). In many actively mobile animals, such as birds, inbreeding may be rendered unlikely simply by the extensive dispersal of the young after they become independent of their parents, making the chance of later meetings with relatives comparatively slight. Sex differences in the degree of fidelity to place also work in this way. In many mammals, the departure of young males (or, in chimpanzees, young females) from the family group makes pairing between close relatives unlikely. In many harem-forming species, such as the savannah zebra and the hamadryas baboon, in which the father remains attached to the family group, the young females, before reaching sexual maturity, are abducted by strange males, thus preventing father-daughter liaisons.

Added to these passive or indirect ways of avoiding incest, cases of active incest avoidance have recently been described. Thus in some bird species sisters form pairs and then show no sexual activity as long as no outside partners are available, thus deterring sexual attention from the father. In wild-living chimpanzees young females have been observed to actively ward off sexual advances from their brothers. The mechanisms underlying active incest avoidance have yet to be elucidated. Certainly individual recognition or kin recognition can be assumed. In primates it appears possible that the subordinate status of a young male with respect to his mother obstructs mother-son pairings. But this cannot be the whole story, for high-ranking males also show restraint toward their mothers (see SOCIAL INHIBITION).

Inclusive fitness An animal's reproductive success in terms of the number of its mature offspring excluding the effects of help or hindrance from social companions, and any increase or decrease in the number of offspring produced by relatives as a consequence of the animal's involvement with them, weighted according to the distance of relationship. By including in the computation of fitness the effect of an animal's activity in perpetuating genes like its own via the reproduction of relatives, along with genetic replication through its own reproduction, it is possible to account for the success of alleles productive of altruistic behavior (see HAMILTON'S RULE; KIN SELECTION). In practice it is difficult to assess the two components of inclusive fitness separately, but it has been argued that a simple tally of the number of an animal's offspring will include any kinship effects.

Two other ways of construing inclusive fitness give higher estimates of reproductive success. The first considers inclusive fitness to be the sum of an individual's own offspring and the offspring produced by its relatives devalued according to the degree of genetic relatedness. This measure has been described as the simple weighted sum; among other objections, it has been criticized for including offspring that the individual had nothing to do with producing. The second conception sums individual fitness and influence on the fitness of relatives, weighted for relatedness. This assessment of individual fitness includes the effects of help or hindrance received from others, which were excluded in Hamilton's (1964) original definition. There is thus some inconsistency in the way the concept of inclusive fitness is used in the sociobiological literature.

Incubation Sitting on eggs to warm them sufficiently for embryonic development to proceed. The term applies mostly to birds and can refer either to the behavior or to the process. In some species, such as passerines, one sex, usually the female does all of the incubating; in other species both members of the pair share more or less equally in the task. Most species develop brood patches, on the ventral surface, which are in contact with the eggs when the bird lowers itself onto them. This facilitates the transfer of heat from the sitting bird to the clutch. In gannets and boobies the inner webs of the feet are applied to the eggs instead of the ventral surface of the body. Megapodes use external sources of heat, such as that generated by mounds of decaying plant material or by volcanic activity.

Incubation is sometimes referred to as brooding, but this term is more often used for the behavior of covering and caring for nestlings;

for the sake of precision, we recommend that brooding be restricted to parental care after hatching.

Incubation patch   See Brood patches

Independent variable   In an experiment, the condition manipulated by the experimenter, who measures its effect on the dependent variable.

Indirect spermatophore transmission   A number of terrestrial animals have evolved ways to effect fertilization without copulation. Lacking copulatory organs, the males of these species convey spermatophores indirectly by depositing them on the ground and then inducing the female to pick them up in such a way as to achieve fertilization. This mode of sperm transmission has evolved independently many times in animals living on relatively moist substrates, where dessication of the spermatophore is unlikely. In vertebrates, it occurs in some species of salamanders, and, among invertebrates, in many arthropods, including myriapods, scorpions, whip scorpions, spiders, mites, and apterygote insects.

The behavior involved in indirect spermatophore transmission is anything but simple or primitive. On the contrary, in many species the behavior patterns appear highly specialized for stimulation of the female and synchronization of her state with that of the male. Examples are the "pairing promenades" of male and female mites, scorpions and pseudoscorpions (see MATING MARCH). Such maneuvers ensure that the female effectively takes up the spermatophore immediately after the male deposits it. Even in species in which there is no direct contact between the mating partners, as in some millipedes, the male may lay down an intricate pattern of signal threads to lead the female to the spermatophore. Only in a very few cases, as in some species of mites, does the male deposit the spermatophore without special behavior patterns or signals to guide the female to it. In these species population density makes it likely that a deposited spermatophore will be picked up by a female without additional measures.

Individual distance The limit to which an animal will tolerate the approach of another. Overstepping this limit leads to an altercation or to withdrawal. Individual distance differs among species, and may fluctuate within a species as a function of time of year or time of day, type of social relationship (as between paired and unpaired animals), sex, and age. In some situations, such as copulation, parental care, and fighting, individual distance may be reduced to zero, even for the most solitary animals.

Individualized group An animal group in which the members know one another individually. Such associations are common only in the more highly evolved vertebrates, principally birds and mammals.

Individual recognition Personal recognition between individuals. As a rule it pertains to conspecifics, but a few interspecific cases are also known. The capacity for individual recognition is most common in vertebrates, especially in birds and mammals, but it has also been reported sporadically among invertebrates, mostly in crustaceans (wood lice, prawns). The partners of a pair, parents and offspring, territorial neighbors, or members of a group may individually recognize each other. Individual recognition is a prerequisite for each kind of social bonding, for the formation of individualized groups, and for the maintenance of rank order. It can be an important means of kin recognition and hence of incest avoidance and of restricting altruism to close relatives. Generally the ability to recognize particular individuals is most important where ties to a specific place, such as a territory or a nest, are insufficient to ensure social selectivity. This is the case, for example, in emperor penguins, who have to pick out their young from the crowd in a creche after returning from foraging at sea.

Individual recognition may depend upon visual characteristics (many birds are able to discriminate among individual head markings), vocalizations (many birds and primates are able to discriminate among individual voice characteristics), or olfactory signals (especially in mammals, fish, and crustaceans). See also PHEROMONE.

Induced ovulation See Ovulation

144

Induction   This term came into ethology via embryology, where it signifies that a particular embryonic tissue can influence the development and differentiation of another tissue. Analogously ethologists sometimes apply the term when, during ontogeny, specific genetic or environmental factors give rise to a new trait, such as a movement or reaction pattern, that would not occur without those factors. Induction, in this sense, is exemplified by a preference formed through imprinting. Induction contrasts with facilitation, which furthers an already established feature instead of inducing a new one.

Infanticide   Infant killing. "Behavior that makes a direct and significant contribution to the immediate death of an embryo or newly hatched [or born] member of the performer's own species" (Mock 1984). Under natural conditions young animals are killed by adult conspecifics mainly in two situations: in adverse or disturbed environmental circumstances, the infant's parents may be the perpetrators; when a new male takes over a harem, he may kill the preceding male's offspring. The first situation has been most often reported in birds of prey, although in many cases it may not be clear whether the parents or other adults actively bring about the deaths of the young, or whether the young die through failure to secure enough food from their parents in competition with their siblings or through some other weakness. The commonest form of infanticide by parents is desertion, but there are numerous instances of physical violence (white storks, long-billed marsh wrens, many rodents). If the killing by parents is followed by cannibalism it is sometimes referred to as 'Cronism' (after the mythical titan who killed and ate his own children). The killing of young by adults other than the parents, with and without cannibalism, as occurs in many gull colonies and in the communally breeding Mexican jay, usually appears to be related to adverse conditions such as food shortage, but not always, so ethologists have sought additional reasons, such as the possibility that internecine killing can serve as a competitive strategy. Sometimes brood or litter size is adjusted to resource supply by the killing of siblings by siblings, variously referred to as Cainism (after the Biblical Cain, who slew his brother Abel), siblicide or sibicide, and fratricide.

The second form of infanticide occurs in harem-forming mammals. When a male takes over a harem after defeating or otherwise replacing the previous possessor, he may kill the unweaned young of his predecessor. As a result the females, which do not come into heat during lactation, will again be ready for mating, and the new male can begin to father his own progeny on them. By removing previous young from the scene, the new male also avoids expending effort in the perpetuation of genes other than his own. This form of infanticide has been described for lions and various primates.

In captivity infanticide occurs more frequently than in the wild, especially as a result of disturbance and abnormally high population densities.

Infantile behavior   The performance of juvenile behavior patterns by adult animals. It is quite common in courtship. Thus in many species of birds the adult females perform begging movements and postures, like those of the young, as part of precopulation behavior. The mature males of many mammals give vocalizations from the juvenile repertoire during the initiation of mating. Begging movements may occur also during courtship feeding. Because infantile behavior tends to reduce hostility, it is a common component of appeasement behavior.

Infant killing   See Infanticide

Infantlike behavior   See Infantile behavior

Information   In the study of animal communication this term is used in two ways: first, in the quantitative sense of cybernetic information theory, equivalent to the notions of uncertainty and negentropy as defined for that context; second, in the more everyday sense, in which information can be classified, restricted, public, sought after in reference books, or stolen by spies. A claim to information in the first sense is open only to the question "How much?"; in the second sense, such a claim might lead to the question "What about?" or the request "Do tell." These two senses are related, but confusion may result if it is not clear which sense applies in a particular instance. Ethologists have to be explicit when they use the term, and readers must be on their guard when they encounter the term unattended by such explicitness.

146

Some ethologists have written of information in developmental and genetic contexts, especially concerning the source of species-specific characteristics. Thus information encoded in genes may be contrasted with formative influences deriving from interaction with environmental conditions. "Information" in this context means something like instructions or specifications for development of a bodily structure or behavioral mechanism. Genetically coded information is sometimes described as a program and as "hard-wired," by analogy with the hardware/software distinction in computer technology. Again, however, one must make sure that genetic and ontogenetic considerations do not get tangled in the kind of confusion that has traditionally dogged the "nature-nurture" issue.

Information theory   Communication theory. Part of the science of cybernetics. The theory was developed in work on the mathematics of signal transmission efficiency in telecommunications systems and then was found to be applicable in the whole field of communication studies, including animal communication. It treats information as a statistically defined quantity whose value depends upon the variety of alternatives from which the signal is drawn and the probability of the signal's occurrence. Amount of information is given by the expression

$$H = \sum_i^n p_i(-\log 2 p_i)$$

where H is the amount of information in bits (BInary digITS), n is the number of options, and p_i is the probability of occurrence of the ith option. This is a precise formal way to express such intuitively obvious considerations as that the word "the" tells us little compared with the word "theodolite" and that chess is harder to learn than checkers. The quantitative treatment of information and communication has refined certain kinds of comparisons of the communication capacities of different kinds of animals, and of the signal capacities of different modes and channels of signal transmission.

Inheritance   See Genome

Inhibition   A term that is applied to a variety of biological phenomena. The two of most concern to ethologists are neural inhibition and behavioral inhibition. In neurophysiology inhibition refers to the suppression of activity of a neuron or group of neurons or an effector organ by influences from other neurons or neuron groups. One such inhibitory relationship exists mutually between antagonistic nerves. It can also obtain between neighboring receptor cells or neighboring sensory neurons, and so lead to a sharpening of contrasts in received stimulus patterns (lateral inhibition: see RECEPTIVE FIELD). Inhibitory processes between different regions of the brain are very important for the organization of behavior. Thus superordinated controls can block subordinated mechanisms and thereby prevent effector organs from operating inopportunely and can determine that the most urgent calls take priority over others, which are also kept from interfering. Cessation of such inhibition can give rise to spontaneous behavior.

In ethology inhibition usually refers to a behavior pattern whose performance is blocked by specific external or internal stimuli or by simultaneous activation of another behavior tendency incompatible with it. A mutual inhibition of this sort has been proposed as a basis for displacement activity (see also TIME SHARING). The inhibition may emanate from another individual. Thus the aggressiveness of conspecifics may be suppressed by appeasement behavior, and certain behavior patterns of high-ranking individuals in a group may block action in low-ranking members (see SOCIAL INHIBITION).

Inhibition of killing   See Social inhibition

Injurious fighting   In contrast to ritualized fighting, a form of fighting in which the contestants not only try to force one another to retreat but also inflict injury, sometimes to the point of causing death (see KILLING BEHAVIOR). As a rule injurious fighting occurs only between members of different species, for example in conflicts between species of similar strength competing for the same resource, or between a predator and a powerful potential prey animal defending itself against attack. Injurious fighting among conspecifics occurs rarely under natural conditions. When it does it is usually between members of different groups in socially living species.

Innate   This is one of the more controversial words in the ethological vocabulary. In one sense "innate" means inborn, hereditary, genetically inherited. In what should be carefully distinguished as another sense, it means ontogenetically developed without any shaping influence from the environment or experience—not learned. In the first sense it means that a specific behavioral characteristic (for example the patterning of a movement, a learning disposition, or discrimination of an object as food or of a species-specific feature as a species marker) is genetically based, which is to say that during the course of phylogenesis information drawn on for its development was accumulated in genetic storage. It does not mean that this characteristic must be present and fully functional at birth or hatching, or that the environment and experience play no part in its realization. Therefore this sense of "innate" does not entail the other. The view of epigenesis is that the development of behavior, like the development of all other features characteristic of a species, involves a continual interplay between what is inherited and what is encountered during growth. In harmony with this, "innate" in the sense of "inborn" has much the same meaning as "congenital" in the phrase "congenital heart failure," a condition that may not occur until relatively late in life, and then only under specific conditions.

Nevertheless, the fallacy of drawing developmental conclusions from facts pertaining to genetic inheritance has been and still is common, causing confusion and misunderstanding on the recurrent issue of "nature versus nurture." Two kinds of question are involved here, and the methods and facts pertaining to the first are different from, and may have little or no direct bearing on, the second. The extent to which environmental interaction or experience is involved in development of a behavioral trait can be investigated by developmental study, not by breeding experiments.

The problem persists because the two questions are closely intertwined, and on occasion inferences about one can reasonably be drawn from premises presented by consideration of the other. For example, if a predatory insect uses perfect prey-catching technique right after emerging from the pupa, the technique could hardly have been learned from prior experience but must have been programmed in the organism. But even with an example like this, heredity can be thought of as setting the limits of developmental potential rather than as predetermining behavior in all its details. Within the range of possibility, environmental influences determine the ways that the genetic information will be acted on to produce a phenotype. For example, the time frame of the sensitive period for imprinting is more or less fixed for a species; that is, heredity dictates at what age the learning can occur, but circumstances determine when it does occur.

The behavioral traits described as innate are as characteristic of a species as are structural and physiological features. Their study can therefore contribute decisively to placing an animal taxonomically, and hence to the construction of a natural classification system.

Innate releasing mechanism (IRM)   See Releasing Mechanism

Innate releasing schema   The historical antecedent (von Uexküll 1957) of the concept of innate releasing mechanism. The term referred to the central (internal) correlate of the stimuli that are effective in eliciting a behavior pattern and was thus posited to account for selective responsiveness. It had more subjective or mentalistic connotations than the later concept and is now obsolete in ethological usage.

Innovation   Inventive behavioral combinations; innovative behavior. New behavioral combinations arising neither from trial and error nor from imitation, but occurring spontaneously on encounter with a novel situation, as though the action sequences anticipate the circumstances and so produce the "right" results the first time. Earlier this phenomenon was referred to as insight learning, but because that description carries a connotation of consciousness it is now rarely used.

In practice one should attribute new behavior to innovation only after excluding the possibilty that the movements occurred by accident or as a consequence of previous experience. Intention movements during an anticipatory phase may indicate what is going on.

Truly innovative behavior is known almost exclusively from primates. The best-known experiments have been carried out with chimpanzees and orangutans. The animals were presented with food out of reach, which could be obtained by combining available objects in certain ways, such as sticks that could be fitted together, or boxes that could be piled on top of one another.

A simple form of innovative behavior found in other mammals, as well as birds and reptiles, is detour behavior; an animal whose path to a goal is obstructed immediately follows an alternative route with success. Such performance implies a certain "insight" into the spatial situation (see COGNITIVE MAP).

Insemination   Introduction of sperm into the reproductive tract or receptacles of a female's body. It is achieved by copulation or indirect spermatophore transmission and leads to fertilization.

Insight learning   Innovation. Problem solving through perception of configurational relationships, as opposed to trial and error. Application of the term to animal behavior derives from Gestalt psychology, in particular Köhler's (1925) study of the intelligence of apes.

Instinct   This is by far the most controversial term in ethology. It was and is used and understood differently by different scientists. To make matters worse, its range of meanings in lay discussions is even wider. Among its many meanings are: purposeful action without "foresight of the ends and without previous education in the performance" (James, 1890); an impulsion to perform some biological function such as migration or reproduction; a stereotyped, species-characteristic action pattern; a genetically determined, endogenously controlled mechanism underlying species-characteristic behavior. Since the term has no unequivocal definition, most ethologists now avoid using it.

When the word does occur in ethology, it usually connotes an inborn behavioral mechanism that manifests itself in an ordered movement sequence, the so-called fixed action pattern, which is activated by specific stimulation of a releasing mechanism controlling discharge of an associated drive. The movements are understood to be endogenously patterned by specific centers in the central nervous system. "Instinctive activity" usually means an ordered sequence of fixed action patterns. The term "instinctive movement," on the other hand, is generally synonymous with "fixed action pattern," so an instinctive activity can consist of a number of instinctive movements. However, these two expressions are not always sharply distinguished from one another. Some ethologists use the terms to mean only that the behavior is species-characteristic and stereotyped, without implying anything at all definite about its ontogeny or causation. Because the terms "instinctive activity," "instinctive

151

movement," and "fixed action pattern" are so loaded with differing conceptual connotations, Barlow (1968) proposed that they be replaced by the more neutral term "modal action pattern."

Tinbergen's (1951) definition of instinct is frequently quoted: "a hierarchically organized nervous mechanism which is susceptible to certain priming, releasing and directing impulses of internal as well as of external origin, and which responds to these impulses by coordinated movements that contribute to the maintenance of the individual and the species." However, he said that this definition was tentative, and he later modified his views about the theory in which it was originally stated.

Instinctive   The adjective shares the range of meaning of the noun. Yet it is still used, usually to convey that the described behavior is genetically inherited and independent of experience in its ontogeny. It is thus more or less synonymous with "innate"; like that word, "instinctive" is often confounded by the mischief of ambiguity, with the genetic and ontogenetic senses assumed to imply one another, despite the lack of logical entailment. Both words should be handled with care, in reading as well as writing.

Instinct model   Models take various forms in science; there are scale models, mathematical models, and various sorts of theoretical models. What they have in common is the representation of one thing as analogous to something else or as though it conformed to some sort of system. In its classical phase ethology developed theoretical models of instinctive motivation, represented graphically by diagrams, that were analogous to hydraulic systems. Lorenz (1950) drew a tank with fluid pouring into it to represent accumulation of "action specific energy" and an outlet controlled by a valve contraption corresponding to the "innate releasing mechanism." Tinbergen (1951) drew a more abstract diagram representing the relationships between superordinated and subordinated motivational centers of his hierarchical model of instinct. However, because the arguments against the concept of instinct apply to these conceptions as well, the instinct models have been more or less relegated to ethology's history.

Instinct-training interlocking A term introduced by Lorenz (1937) (originally "drive-training interlocking") to refer to reciprocity between genome and experience in the development of a behavioral characteristic (see ONTOGENESIS), as a consequence of which innate and acquired components mesh. It is as though a genetically programmed behavioral sequence is interspersed with empty spaces that must be filled by interaction with the environment. The position of the gaps is genetically determined, but their content—the kind of information entered—is dependent on experience. Examples that fit the idea of instinct-training interlocking are the killing behavior (neck bite) of European polecats and the opening of nuts by squirrels. In both behavioral sequences the forms of the component movements are independent of experience, but their orientation has to be learned.

Instrumental conditioning   See Conditioning

Intelligence   This word has been applied to animal behavior in various ways. Some writers consider it more or less equivalent to learning ability, others reserve it for evidence of reasoning, inventiveness, or insight. Perhaps the majority view is that, even in human beings, intelligence is no one thing, but a collection of capacities, including perceptiveness, imagination, memory, mental and behavioral flexibility, problem-solving ability, and ability to assess situations in the light of past experience. The validity of single measures of intelligence has been questioned, as has the extent to which the results of tests of intelligence can be compared across groups of people of different cultures. The difficulties of measuring and testing intelligence in comparable ways across different animal species are even greater. Consequently comparisons of animal intelligence need to be specific about what has been measured and tested, and should consider whether differences may result from perceptual or motor limitations rather than cognitive capacities.

It follows from this view of intelligence as multifaceted that interest in the adaptive significance and evolution of intelligence must take into account more than a single capacity.

Intentionality   In everyday language, having intentions—planning or undertaking action in pursuit of subjectively held aims or goals.

However, philosophers of mind use the term in a broader sense to refer to propositional attitudes: all mental states that refer to a content of some sort—that are about something. Examples are believing, wanting, hoping, knowing, and understanding. One cannot believe purely and simply, one must believe *in* something or have a belief *about* something. In contrast, it is possible to ache or shiver without referring to anything by so doing. The intentional idioms have some logical peculiarities compared with terms applying to physical states and objects; "intensional" logic (note the difference in spelling) is differentiated from "extensional" logic. For example, a statement containing an intentional idiom may not remain true if a different designation of the same object is substituted; in contrast, the truth of extensional statements is unaffected by such substitution. Thus it was true both that Hamlet intended to kill the man behind the arras and that the man behind the arras was Polonius; this does not entail that Hamlet intended to kill Polonius, but does entail that the man behind the arras was Ophelia's father. (This nonsubstitutability of codesignating terms is referred to as referential opacity.) Some philosophers have argued that intentionality in this broad sense is "the mark of the mental" (Brentano, 1874).

The mentalistic connotations of intentionality, in both its narrow and broad senses, have persuaded most ethologists to have nothing to do with it. However, the recent emergence of cognitive ethology has made the possibility of intentionality in animals a matter to be taken seriously. For example, ways have been explored for determining whether a vervet monkey giving an alarm call does so with the intent of informing companions that a predator of a specific sort is in the vicinity, and an experimental approach has been devised for testing whether a bird "playing the broken wing" is deliberately attempting to deceive the predator toward which it directs the behavior (see DISTRACTION DISPLAY). Such cases illustrate that considering an animal's behavior in terms of intentionality can lead to productive questions and useful observations that might not otherwise occur to behavioral scientists.

Intention movement Performance of the initial parts of a behavior pattern without the rest of the pattern. Movement that stops short of its full course. For example, if a bird makes repeated pecking movements prior to making a full nest-building movement, the pecking is viewed as intention nest-building. It is an expression of the animal's motivational state and can therefore serve to communicate to conspecifics what the animal is primed to do next. Thus the threat behavior of many animals consists of intention movements of actual fighting actions, such as biting or pecking. In a number of bird species an individual's intention movements of takeoff are stimulating to the members of a flock at rest, rousing their tendency to fly and resulting in all taking wing together (see SOCIAL FACILITATION). Frequently the effectiveness of intention movements as a signal has been enhanced in the course of phylogenetic history by ritualization. Intention movements are thus one of the main evolutionary sources of display behavior.

It should go without saying that "intention" here does not imply subjective intent (see INTENTIONALITY).

Interaction This term is used in several ways in ethology. Most often it refers to an encounter between animals in which each responds to the presence of the other. Signals may be exchanged, or one animal may withdraw at the other's approach; such mutual stimulation and responding as occur in courtship, agonistic behavior, parental care, and so forth exemplify interaction in this sense. An animal is sometimes said to show interaction with its environment, in which case the influence is usually more one-sided; the environment affects the animal more than the animal affects the environment. In statistical analyses a third and more technical usage is current: in an analysis of variance a portion of the variance may be due to the joint effects of two or more factors over and above that due to their independent effects, and this is called an interaction.

Internal clock Endogenous clock; internal chronometer; time sense. The capacity to tell time independently of external signs such as the position of the sun. This ability depends on the fact that certain physiological processes recur at regular intervals (see PERIODICITY). The internal clock plays an essential role in the language of bees (see DANCE LANGUAGE) and in the navigation of migratory birds and other animals.

The actual seat and physiological basis of endogenous rhythmicity have yet to be pinned down. In vertebrates recent investigations make it increasingly evident that the pineal organ plays an important part in the process. This vesicular structure is at the tip of the pineal gland, a tear-shaped attachment to the roof of the midbrain.

Internal stimulus Sensory stimulation from within an animal's body, such as that caused by a distended bladder. It is registered by interoceptors.

Interneuron See Reflex arc

Internuncial neuron See Reflex arc

Interoceptor Also sometimes called an endoreceptor. A sense receptor that registers the state or change of state within an individual's body. Interoceptors register such variables as blood pressure, body temperature, degree of fullness of the stomach, hydration state, oxygen debt, and acidity of the blood. Interoceptors in muscles, tendons, and joint tissues that tell of the body's position, movement, and strain are referred to as proprioceptors.

Interspecific releaser A releaser that serves for communication between members of different species. An acoustic example is the alarm call of many songbirds, which is also understood by other species (see WARNING BEHAVIOR). Interspecific releasers also occur in chemical communication, as occurs with so-called ant guests. These partly harmless, partly destructive insects (beetles, other ant species, silverfish) secrete a substance that attracts the host ants and elicits, say, parental care behavior (feeding or transport to safer ground in times of danger). The substance emitted by beetle larvae apparently simulates the pheromone of the larvae of the host ant and can therefore be described as a false pheromone (see STIMULUS MODEL). In many cases it appears to be even more effective than what it mimics, and so presents a natural supernormal stimulus.

An interspecific tactile signal is given by gobies to the shrimp with which they live in symbiosis: as soon as a predatory fish approaches the shrimp's hole, the goby gives a warning signal by sharply beating its tail, which the shrimp registers with its long antennae and responds to by retreating into its hole.

156

Interspecific territoriality Territorial behavior between members of different species. The main (proximate) function of a territory is the spacing out of competitors. The most direct competitors confronting an individual are its conspecifics, because the environmental requirements (food, breeding sites) are much the same for all. Therefore, as a rule, territories are defended only against members of the same species. However, when two species living in the same region have closely similar ecological requirements and therefore compete with one another quite directly, it is advantageous for an individual not to tolerate members of the other species in its territory. As a consequence both species show convergence in appearance, vocalization, and specific movement patterns. This allows for mutually understood signaling that serves as an interspecific releaser. Interspecific territoriality has been described for frogs, salamanders, lizards, some mammals (rodents, shrews) and for a large number of bird species, including woodpeckers, hummingbirds, honey eaters, warblers, shrikes, goldcrests, and corvids.

Intervening variable When an independent variable and a dependent variable can be related to one another quantitatively by some coefficient, this latter is referred to as an intervening variable. Defined in this way, the term extends no farther than the work it does in the equation relating the results of the experiment. However, "intervening variable" has sometimes been used to mean hypothetical construct—some process or state or entity that would account for the relationship in terms of an underlying mechanism, for example, level of activation of a center in the central nervous system. The term "drive" has been used both as an intervening variable and as a hypothetical construct.

Intimidation Effective threat; the cowing or daunting of one animal by another. Thus visual or vocal signals and scent marking can halt the approach or cause the retreat of a territorial trespasser; in groups structured by a social hierarchy, the mere presence of a dominant animal can deter the activity of a subordinate. See RANK ORDER.

Intraspecific behavior Behavior having to do with conspecifics. Interactions between members of the same species. The concept is close to but not completely interchangeable with social behavior; for one thing, in some cases interspecific interactions can be regarded as social in nature.

Intrinsic selection   See Selection

Introspection   Attending to the content of one's own experience. By providing a foundation for comparison, introspection can be an important means of understanding the behavior of other people, as when we achieve empathy with someone else's feelings. The term used to be used similarly in comparative psychology as a means of trying to understand the mental lives of animals (see ANIMAL PSYCHOLOGY). Now, however, it is generally excluded as a method for the objective study of animal behavior because the results of generalizing from our own introspection to the experiences of animals are unverifiable in any scientific sense, the minds of animals being inaccessible to direct investigation.

Inventive behavioral combinations   See Innovation

Investment   See Parental investment

IRM   See Releasing mechanism

Irreversibility   A term used especially in the early literature on imprinting for the marked persistence and stability of object preferences established during the sensitive period. Irreversibility is most notable in sexual imprinting, in which sexual preference is determined by a brief period of elicitation of the following response. While most cases of sexual imprinting appear to be irreversible, many other imprinting-like processes are later alterable, although only to a certain degree, and sometimes—as with the separation syndrome—only after great effort. In these cases the less absolute terms "durability" or "stability" are more apt. Enduring effect has also been found for some instances of song imprinting, place imprinting, habitat imprinting, and host imprinting.

The biological significance of extensive stability lies in the function of information storage: it ensures that biologically relevant information acquired during the sensitive period—for example, about the

158

characteristics of an animal's own species or adequate breeding habitat—is not preempted or replaced by adventitious false experience when the animal leaves its parents and comes in contact with members of other species or enters a region unsuitable for breeding.

Irritability   The capacity of an organism to react to conditions impinging on it in adaptive and reversible ways; capacity to respond to a stimulus. Irritability is one of the fundamental characteristics of life. It consists of receptivity, conduction, and action (movement or secretory activity), capacities that exist to some degree in all living tissue, from single cells, which may be specialized (receptor cells, nerve cells, muscular and gland cells), to whole multicellular organisms.

Isolating mechanism   "Biological properties of individuals that prevent the interbreeding of populations that are actually or potentially sympatric" (Mayr 1963). The mechanisms are divided between those that act prior to pairing, and so prevent crossing between members of different species or populations from the start, and those that take effect only after hybrid pairing, and so upset or prevent the results of such pairing (for example, death of the fertilized ovum). For the first kind of isolation mechanism, behavioral characteristics are very important. See REPRODUCTIVE ISOLATION.

Isolation experiment   Also referred to as experience deprivation and deprivation experiment. Deliberate withholding of certain possibilities of experience in the rearing of an animal. With this method one can determine which capabilities develop normally, even in the absence of what might have appeared relevant experience, and what deficits go with what kinds of deprivation. Evidence from isolation experiments has been the basis of inferences about which components of a species' behavior are innate, but this is a controversial matter. Inference from an experiment with experience deprivation is possible in only one direction: if the behavior of a deprived animal turns out to be normal, one can conclude that the experience withheld is not necessary to development of that behavior. But if development is abnormal, the reverse conclusion does not necessarily follow, because the deficits could also be the result of unnatural features in the circumstances of the isolation experiment.

Experience deprivation can be effected by rearing animals in seclusion (say, alone in soundproof boxes) or by tampering with sensory or motor function, (by sealing the eyes, deafening, or deafferentation, which effect sensory deprivation, or by restraining use of the limbs or using paralytic drugs, which effect motor deprivation). As a rule only one kind of experience is withheld, such as deprivation of a specific sensory capacity or of a specific kind of object (prey, nest material). Restricting contact with conspecifics—mother, parents, siblings, or unrelated peers—is called social deprivation. Animals raised under conditions of experience deprivation may be referred to as Kasper-Hauser animals, especially when the degree of deprivation is extensive.

Experience deprivation is especially favored in bioacoustical research. The withholding of specific acoustic models (vocalizations of conspecifics) can determine which vocalizations are acquired independently of experience and which require specific models for their development (see SONG; TEMPLATE). The complete shutting out of all acoustic models is possible in special sound-isolation chambers or rooms. Distortions of behavioral development resulting from experience deprivation are referred to as deprivation syndrome.

J U V E N I L E

J

Juvenile   Subadult animals, not yet sexually mature. Sometimes the term is used like "adolescent" to refer to the developmental stage between infancy and adulthood.

Juvenile characteristics   The specific characteristics or character combinations that distinguish young animals from the adults of their species. The term is used mainly for the appearance of the young animals, such as juvenile plumage in birds, but it can also refer to behavior and vocalization.

Most juvenile appearance is inconspicuous and probably serves as a first line of defense against predators (see CAMOUFLAGE). In addition, juvenile characteristics may play a role in social behavior in two ways: first, they may present releasers for parental care behavior, as is the case with the conspicuous palate markings of many young birds; second, and contrastingly, they may serve for social shielding, that is, for preventing adverse social attention before the development of sexual and agonistic releasers and of behavioral competence in their use. This latter function is especially important in species with long-lasting parental care and in group-living species in which the young enjoy a certain "freedom to play" before they are integrated into the social hierarchy. Juvenile dress plays an important part in the highly developed social relations of many primates, for the young animals give signals that can inhibit aggression from adults (see AGONISTIC BUFFERING). In these species, as a rule, the appearance of the young is very different from that of the mature animals, and may even include quite conspicuous coloration (so-called baby coloration). For example, infant black colobus monkeys are almost completely white, in contrast to the predominantly black coloration of the adults.

Juvenile song   See Subsong

KIN RECOGNITION

K

Kaspar-Hauser An animal reared under conditions of severe experience deprivation and hence denied opportunity for the learning necessary for normal behavioral development (see ISOLATION EXPERIMENT). It is virtually impossible to contrive a total Kaspar-Hauser animal, since even when reared in complete isolation in lightproof and soundproof conditions, an animal still has available to it certain kinds of experience, at least of its own body. Depending upon the question being investigated, the young animal is often deprived of only certain sorts of stimulation, such as social contact with conspecifics. Animals treated in this way are sometimes described as Kaspar-Hausers of the second degree or partial Kaspar-Hausers.

The designation originates from a foundling child of that name, who appeared suddenly in Nuremberg in 1882 and caused a widespread and considerable sensation. There is an extensive literature on his subsequent life, but his origin remains a mystery. All that he could remember was that he had lived in a dark room. His mental development remained severely retarded all his life as a result of the impoverishment of experience during childhood. Hence the parallel to the experimental contriving of Kaspar-Hausers in developmental research.

Key stimulus   See Sign stimulus

Killing behavior   Killing occurs not only in hunting by predatory species but also in intraspecific interactions, where adults as well as young may be killed. Killing among adult conspecifics was, until recently, considered to be unnatural—a result, for example, of closed escape routes in the confines of captivity. However, fighting with fatal outcome has been observed under natural conditions in an increasing number of birds and mammals. Even so, most of these cases occur in exceptional circumstances, such as a shortage of resources (food or breeding territories) caused by very high population density. In group-living mammals the victims are usually outsiders. Evidently inhibitory mechanisms prevent or limit the use of lethal aggression between members of the same group (see APPEASEMENT BEHAVIOR; SOCIAL INHIBITION).

Kinesics   See Nonverbal communication

Kinesis   Movement that takes an animal into favorable regions of its environment and away from unfavorable conditions; it is not governed by orientation and is therefore not a taxis. There are two forms. In the first, orthokinesis, the rate of locomotion varies with the stimulation: when conditions are adverse the animal races ahead; as they improve, it slows down; when they are optimal, it marks time. In the second form, klinokinesis, the rate and magnitude of change of direction are functions of stimulation: when conditions are adverse the animal switches direction repeatedly and through wide angles; when a turn carries it to better conditions, it continues in that direction until conditions start to deteriorate, at which point it changes direction again. This mode of locating and keeping to favorable conditions was earlier described as a form of trial-and-error behavior and as the avoiding reaction. The two forms of kinesis probably go together in most cases. Kinesis is the only possibility for animals that lack sensory equipment for discriminating the direction of stimuli, such as ciliate protozoans (*paramecium*), which provided the classic example of the avoiding reaction.

Kinesthesia   Sensations of movement and body position. An animal's perception of its own movement and posture. The sensations derive from proprioceptors in the joints, muscles, and sinews, and, in vertebrates, from the membranous labyrinth of the inner ear, where there are sensors of angular and linear motion in the semicircular canals and sensors of static position with respect to gravity in the otolith organs of the vestibular region (utricle and sacculus). Kinesthesia can constitute a kind of motor memory, which makes possible a more or less automatic repetition of movement patterns.

Kinesthetic learning can be instilled by a form of training in which the trainer actively forces the animal through specific movements or into specific positions (so-called putting-through method), as in the teaching of sign language to chimpanzees.

Kin recognition An animal's ability to differentiate between relatives and nonrelated conspecifics. It is known to occur in higher vertebrates, especially mammals and birds, and also sporadically in invertebrates (desert woodlice, honeybees). It may include personal recognition of single individuals; collective recognition of members of a genetically related group by some mechanism such as group odor or hive odor; or phenotype matching, in which an animal distinguishes between others on the basis of similarity to itself.

Kin recognition can contribute to avoidance of mating between close relatives and, conversely, may possibly lead to a bias in favor of mating between (distant) relatives. In addition it allows altruistic behavior to be restricted to close kin (see KIN SELECTION).

Kin selection Within a population, increase of genes that cause individuals to promote the survival and reproductive success of relatives. An individual has some genes in common with its relatives, so their reproductive success will tend to perpetuate those genes, to a degree that correlates with the relatedness (see KIN SELECTION THEORY). The conditions for kin selection are expressed in Hamilton's rule: $rb - c > 0$ where r is relatedness, b is benefit to recipient (added number of surviving offspring) and c is cost to the helper (forfeited number of surviving offspring). The selection is naturally greatest for closest relatives; hence parental care has independently evolved a great many times. Some biologists exclude from kin selection the effort an individual puts into promoting survival of its own offspring, presumably because helping one's sister to care for her baby is of a different order from caring for one's own, as far as inclusive fitness is concerned. There is thus some variation in usage, even in sociobiology, where the concept of kin selection was first formulated.

Kin selection theory   A theory developed in sociobiology for explaining the evolution of altruism. The occurrence of altruistic behavior appeared inconsistent with natural selection, which is supposed to favor characteristics that benefit the individual in the struggle for existence, that is, characteristics that conduce to selfishness. The inconsistency is resolved by considering that an individual's actions on behalf of relatives can perpetuate genes that are the same as the individual's in proportion to the degree of relatedness. Thus an individual has half its genome in common with its sons and daughters, a quarter with its grandchildren, and an eighth with its great-grandchildren. Siblings have half their genes in common (for identical twins the genome is the same for both), half-siblings have a quarter, and first cousins have an eighth. With regard to altruistic behavior, this sharing of genes between relatives allows for an estimate of genetic compensation. Even if the individual dies without reproducing, its genome will be represented in the next generation just as it would have from reproduction if, for every offspring it might have had, two grandchildren, two nieces or nephews, four great-grandchildren, or four first cousins survive. An individual's total genetic legacy to a gene pool can thus be divided between the results of its own reproduction, discounting any help or hindrance received from others, and the results of any help or hindrance it gives to relatives, the latter weighted according to the degree of relatedness. These two contributions together constitute the individual's inclusive fitness.

Among other things, this theory has provided explanation for the existence of sterile worker castes in the socal Hymenoptera (ants, bees, and wasps) where the peculiar haplodiploid mode of sex determination makes worker females 25 percent closer to their sisters, genetically, than they would be to their daughters. It also accounts for the numerous instances of nepotism—animals putting themselves at risk or expending effort only for close relatives and not for conspecifics at large.

Kleptobiosis   See Kleptoparasitism

Kleptogamy   "Sneaky mating." A relatively new term for copulations "stolen" by males lacking a territory, subordinate in rank, or not belonging to the group or pair (see MULTIMALE GROUPS; SATELLITE MALES). It has been described in a number of vertebrate species and some insects. For example, determinations of paternity among rhesus monkeys living in virtually natural conditions showed that more than 5 percent of the offspring came from males that did not belong to a group. In bullfrogs the smaller males sit quietly near larger croaking males and may try to intercept females attracted by the display. Young male elephant seals may try to inveigle their way into a harem by acting like females. In blue-gilled sunfish, males take three forms: quite conspicuously colored parental males, which defend territories and construct nests for females to lay their eggs in; inconspicuously colored males, which are neither territorial nor parental and which sneak into nests and discharge semen after females have spawned their; and males that mimic females in appearance and behavior and gain access to the nest in that way. In many species, at the times when copulation has the highest probability of achieving fertilization, the males show increased vigilance toward their females, which evidently serves to impede or prevent kleptogamy (see GUARDING BEHAVIOR).

Kleptoparasitism   Food robbing. The appropriation of commodities (food, nest material) accumulated by members of another species. Tropical frigate birds live mainly on food they get by harassing seabirds, especially terns and boobies. Many types of kleptoparasites live in the colonies of social insects. For example "thieving ants" take up residence in the nests of other ants and live on their food stocks (see SOCIAL PARASITISM). Occasionally, as in terns, food robbing has been observed among members of the same species. Some entomologists use the term "kleptoparasitism" more narrowly for the behavior of females that raid the provisions of others, usually of other species, to get food for their own young; they distinguish this from kleptobiosis, the robbing of food stores or scavenging by one species in the nests of another, but without living in close association with that species the way a parasite lives with its host.

Klinokinesis   See Kinesis

Klinotaxis   See Taxis

K-selection   A form of selection typical of conditions of ecological stability where the population size is at a stable maximum. K-selection favors the production of relatively few offspring that develop slowly and are carefully nurtured, so as to be competitive with others of their generation. Thus it promotes social systems that involve such features as pair bonding and parental care. Compare r-SELECTION.

LEK

L

Labyrinth In anatomy, part of the inner ear of vertebrates, consisting of the semicircular canals and the otolith organs (saccule and utricle). The structure, also called the membranous labyrinth and the vestibular organ, contains mechanoreceptors, which register position and movement of the body: tilting and turning, acceleration and deceleration. See ORIENTATION.

In animal experiments the term is synonymous with "maze."

Language Verbal expression. A system of communication consisting of words and grammatical rules. When uttered, language constitutes speech. From this root meaning the reference of the word has been extended figuratively or by analogy to animal communication, though mostly in nonscientific writing describing "animal speech" or "animal language."

The extent to which human language and various kinds of animal communication might be comparable is a matter of considerable interest and controversy. Some people take the view that where symbolic meaning can be attributed to animal signals—that is, when the signals are used to refer to something in the outside world and are not merely symptomatic of an animal's emotional or motivational state—it is legitimate to speak of them as speech or language. In this sense language can include nonvocal modes of signaling. The most remarkable form of nonverbal language that has evolved in animals is the dance language of honeybees. Worker bees returning to the hive from foraging trips can inform their sisters about the direction, distance, and quality of a food source through movement sequences called dances (round dance, waggle dance). This mode of communication has the unusual feature in common with human language that it can convey information even in the absence of the objects or places to which it pertains. However, in other respects there are profound differences. The bee language is a rigid code narrowly restricted in its range of possible reference, in contrast with the flexibility and "openness" of human language. Moreover, human language depends upon learning and tradition for individual acquisition and transmission from generation to generation to a degree that has no parallel in the development and inheritance of the communication behavior of bees.

The closest approach to human language learning by nonhumans has been achieved in the teaching of human sign systems to chimpanzees. The apes so taught have acquired a "vocabulary" of over a hundred "words," which they can use and comprehend, often in sentencelike combinations and sometimes in novel but still meaningful ways. Although there is some doubt and controversy about whether these apes have acquired rules of grammar, the ape language projects have led to a general revision of opinion about the cognitive capacities of these animals.

Contrasting with true speech are the words and sentence fragments that some birds are capable of learning and uttering. These imitations lack symbolic character. In some cases they are instrumental (operant) responses reinforced by the person taking care of the birds. In other cases they are the equivalent of calls exchanged between conspecifics, learned by mimicking the sounds of a model, such as a territorial neighbor. For caged birds the only model available is the caretaker.

Larva A free-living juvenile stage of an organism that differs substantially from the sexually mature adult in form, way of life, possession of special structures (larval organs), and lack of adult structures such as sex organs. Profound differences in behavior may distinguish larval from adult stages, including mode of feeding, locomotion, orientation, and social patterns. The transformation of larvae into adult forms is called metamorphosis. In the holometabolic insects this involves an intervening pupal stage, inside a special case or chrysalis in many instances, during which the body is extensively reconstructed. Familiar larval forms are the tadpoles of batrachians, the caterpillars of moths and butterflies, and the maggots of flies.

Latency   Reaction time. The time interval between occurrence of a stimulus and occurrence of response to it (a behavior pattern or a nerve impulse, say). Also, in contexts of learning, the time between a conditional and an unconditional stimulus (light, electric shock), or between performance of an operant response and reinforcement. See CONDITIONING.

Latent learning   When an animal is exposed to a situation with no immediate beneficial or harmful contingency and is later able to use that experience in obtaining reward or avoiding hurt, it is said to show latent learning. Thus rats allowed prior access to a maze mastered it more quickly than rats new to the maze when arrival at the goal box was reinforced. See COGNITIVE MAP.

Lateral inhibition   See Receptive field

Law of heterogeneous summation   See Stimulus summation

Learning   Behavioral change effected by experience, including formation of habits by conditioning and acquisition of information that gets stored in memory. In ethology the term covers all processes that lead an individual to adapt its behavior to prevailing circumstances, as well as the adaptations themselves, that is, all changes in behavior that are a consequence of individual experience. Learning can be defined "in the widest sense as an adaptive modification of behavior" (Lorenz 1981) in the life of an animal. Learning processes have been considered to include four phases: stimulation, recording, storage, and retrieval: the registering of environmental input by sensory receptors, entry of the information into central nervous structures, retention of the information, and its recovery (see CELL ASSEMBLY;

ENGRAM). Some learning theorists have supposed that association, either of stimuli and stimuli or of stimuli and responses, and reinforcement are also essential to the process. However, the emphasis on information may suggest more than is needed to account for relatively simple forms of learning such as habituation (which also relates awkwardly to assumptions of association and reinforcement), and information has a cognitive connotation irrelevant to the acquisition of skill through practice. Also the passivity of information storage tends to divert attention from the ability of animals to actively generate their own experience. And reference to the central nervous system seems to imply that animals lacking a central nervous system, such as protozoans, are incapable of learning. In reality learning of some sort appears to occur in all multicellular animals, and probably in many unicellular (or acellular) animals as well.

Behavioristic learning theory, which derives historically from associationistic psychology, has traditionally assumed that throughout the animal kingdom learning conforms to a small set of principles, the universal laws of learning. Ethologists and some comparative psychologists question this assumption, arguing that although learning can be used as an "achievement word" to cover experience-induced behavioral change in general, the processes underlying such change in animals of different kinds may well be profoundly different in both mechanism and principle. There is good reason to think that an ant's tiny brain and a rat's relatively large brain must process the effects of experience quite differently, even where their performances are comparable, as in maze learning.

A distinction can be drawn between obligate learning, which is essential to the life of the animal, and facultative learning, which is not. Examples of the first are all imprinting processes and learning in connection with predator defense and obtaining food. Examples of the second are individual recognition and learning derived from exploration and play behavior. However, as these examples illustrate, the distinction is not hard and fast; the importance of a particular kind of learning for survival and reproductive success is a matter of degree rather than a difference in kind.

Learning capacity   The ranges and limits of what, when, and how an animal can learn; that is, the kinds of experience it can register and store; the kinds of behavior patterns it can acquire; the associations of stimuli and responses it is capable of making; its ability to retain the effects of experience, to unlearn associations, and to replace them with new ones; and the quantity of learned information it can store at any one time. The comparative study of learning thus includes consideration of sensory capacities and motor capacities, as well as the intervening processes and processors between input and output.

In general, learning capacities are of greater range and sophistication in phylogenetically more recent ("more highly evolved") groups than in phylogenetically older groups. But there are many exceptions to this, especially in quite specific learning dispositions. In closely related species the larger animals often have greater learning capacity than the smaller, at least in some respects. Aspects of an animal's learning abilities can change during the course of its life, as is well understood in the case of humans. An extreme example is that of imprinting, for the same experience may have quite a different effect on the animal, depending on whether it occurs within the sensitive period or outside it.

The concept of learning capacity, like that of learning disposition, is comparable to the notion of "reaction norm" in genetics. That is to say, it implies a genetically determined range within which interaction with environmental factors affects the formation of a specific character.

Learning disposition Predetermined, presumably genetic, constraints on learning capacity, which limits the range of things an animal can learn and the ease with which it can learn within that range. The notion of learning disposition draws attention to the genetic substrate underlying any animal's characteristic learning capability. Also, and more specifically, it concerns the observation that many animals learn certain things very quickly. These are almost always of crucial biological significance in their lives, such as the characteristics of food and predators, while less significant things are learned slowly or not at all. Such predisposition to learn vitally relevant things more readily than anything else has been described in recent learning literature as a dimension of preparedness, which contradicts the equipotentiality assumption implicit in earlier learning theory, according to which one stimulus-response association is as easily formed as another. Learning dispositions vary with the ecological adaptations of different species, and even of varieties within a species. Also the readiness to learn can change in the course of ontogeny; the most marked contrast is that between the sensitive period of imprinting and later stages of development. Many socially living animals are predisposed toward learning to recognize members of their group individually. Male songbirds usually learn only the song of their own species, which they select from the diversity of songs they are exposed to; their preference for a certain tone quality or note pattern is prior to and independent of experience. Correlation of learning disposition with selection pressure is illustrated by comparison of two races of honeybees. Under identical testing conditions, bees from the Carniolan race of northwest Yugoslavia were much better at learning landmarks that served as orientation guides in finding food than were bees of the Italian race. This difference in learning disposition makes biological sense; the Italian bees have such good weather conditions that they can rely exclusively on sun-compass orientation, but the Carniolan bees live in a region of inclement weather and much of the time must use features of the landscape to orient themselves.

Lek   A communal mating area within which males hold small territories, which they use solely for courtship and copulation (see COMMUNAL COURTSHIP). Females are attracted to a lek by the displays of the males and then choose mating partners from among them, usually favoring males with territories in the center. As a consequence one or a few of the males do most of the mating. It is thought that the more peripheral males benefit by being genetically related to the central males, whose chances of breeding they promote by adding to the communal display (see KIN SELECTION). They may get an opportunity to mate either by snatching a female when a central male is occupied with territorial defense or by successfully competing with a central male and taking over its spot. It is unclear how this peculiar kind of promiscuous mating system has evolved, as it has numerous times, for example, in the picture-wing *Drosophila* of Hawaii; in ten families of birds, including hummingbirds, manakins (Pipridae), sage grouse, wild turkeys, ruffs, and bowerbirds; in hammer-headed bats; and in the Uganda kob. In some cases the lek is a specially prepared or adorned arena. For example, the South American manakins clear an area of leaves on the forest floor, and the bowerbirds of Australia and New Guinea construct passageways, platforms, or grottoes, which they decorate with flowers, shells, stones, and leaves.

Lesion   In ethology and physiology, deliberate destruction of neural tissue for experimental purposes, as by surgical severing of a nerve tract in the brain. Refinements in technique have made it possible to confine destruction to single neurons and to use selectively binding substances to effect chemical lesioning. Lesioning studies have contributed a great deal to understanding of the functional anatomy and physiology of nervous sytems, but because they usually result in loss of function, they need to be complemented by experiments using neural stimulation techniques.

Lloyd Morgan's canon   "In no case may we interpret an action as the outcome of the exercise of a higher psychical faculty, if it can be interpreted as the outcome of one which stands lower in the psychological scale" (Lloyd Morgan 1894). In reaction largely to the anthropomorphism of his contemporaries in their interpretations of animal behavior, Lloyd Morgan took a tough-minded stance, that of adopting the minimum postulate sufficient to account for an animal's actions. Behaviorists then applied this notion with even more reductive force than Lloyd Morgan. His canon is still generally followed, even though there is no guarantee that this is the simplest of all possible worlds as far as control of behavior is concerned.

Local enhancement   An animal's directing of attention toward some object or place in its environment as a consequence of observing how others act with regard to that object or place. Thus many animals find their way to a source of food more quickly if they see others looking for that place. Like social facilitation, local enhancement is to be distinguished from imitation. Each animal finds its own way by trial and error, even though the process is facilitated by observing the other animals.

Locality imprinting   See Place imprinting

Locomotion   Movement from place to place; passage. Change of place by a freely moving organism by its own power, as by crawling, running, climbing, swimming, or flying. Not included in locomotion are position-changing movements of sedentary animals, such as the tentacle waving of sea anemones and hole- or tube-living worms, and passive transport by wind or water current. The motor patterns used for locomotion vary widely among different kinds of animals, and even for a particular animal the patterns may vary with speed, as the gait of a horse goes from a walk to a canter to a full gallop.

Lone wolf A term usually used to describe a single individual of a social species, as an old bull in a mammal group such as elephants, that detaches itself more or less permanently from the herd. Sometimes, however, solitary animals are described as lone wolves.

Long-term memory See Memory

Long-term pair bond Durable monogamy, or pair bonding that transcends a single reproductive period. In avian species with long-term pair bonding, pair formation may precede reproductive maturation. The description "lifelong" for protracted cohabitation should be restricted to those species (some raptors, gray geese) for which it has been shown that pair bonds endure virtually for the lives of the partners. (On the problem of using the term "marriage" in ethological contexts, see that entry.)

Lordosis Also referred to as the lordosis reflex. The posture adopted by many female mammals before and during copulation. It consists of concave arching of the back, sideways spreading and extension of the hind legs, lifting of the tail, and forward and downward pointing of the head. It occurs only when a female is in estrus, when it can be elicited by tactile stimulation of her flanks.

Lying-out period Used of ungulates. In many species of antelopes and deer there is a period soon after birth when, after being suckled, the calf or fawn actively leaves its mother's side and lies down in cover, remaining thus couched and concealed until called by the mother for its next feeding. Among the species showing this behavior are oryx antelopes, kudu, many gazelles, and red deer. Lying-out provides concealment from predators during the time before the young animal is able to run and flee, but it requires a habitat with adequate cover and a social organization (a stable home range or territorial system) that allows the mother to easily rejoin the herd with her young at the end of the lying-out period.

In some species, such as the dikdik, the mother leads the calf to the hiding place after suckling and induces it to lie down. In species such as roe deer, where twins or triplets are the rule, each young chooses its own lying-out place some distance from the others. These precocial offspring are sometimes collectively referred to as hiders. They contrast with followers in that they spend the greater part of the day apart from their mothers. However the distinction between hiders and followers is not hard and fast: many intermediate cases are difficult to classify as either, and the young of even the most extreme hiders eventually make a transition, gradual rather than abrupt, to following the mother.

MOUTH BREEDER

Macrosomatic animals Olfactory animals. Animals with a well-developed sense of smell. The division into macrosomatic and microsomatic species is used chiefly for mammals, most of which—in contrast to birds—possess sharp noses.

Male care A term that has come into increasing use in recent mammalian literature, especially on primates, for parental care performed by males. In the majority of birds and many fish, males contribute about a half share to raising the young, or even do the whole job alone. In contrast, in mammals the females are "automatically" consigned to the bearing and suckling of the young, and male parental behavior is rare. The extreme exception is presented by marmosets, in which the males take on practically all the parental care apart from suckling. In other mammals male parental involvement consists mainly in defense of the young, social grooming, social play, and, in primates, occasional holding and carrying of the young. At least in primates, male parental care apparently serves an important function by providing learning experience in, for example, sex differences in behavior (see ROLE). In multimale groups the males appear to concentrate parental care on their own offspring. Cases in which it has been established that the parental behavior is restricted to a male's own young are referred to as instances of paternal care.

Because of the difficulty of clearly distinguishing between parental care and other kinds of behavior, the concept of male care is sometimes extended to include the totality of behavioral interaction between males and young animals, with the exception of aggressive encounters.

Manipulation   Animals can effectively handle parts of their environment, and they can effectively handle one another. Environmental manipulation includes nest building, burrowing, nut cracking, tool use, and numerous other activities. It has been argued that animal signaling can be construed as a form of manipulation—a means by which an animal may try to maneuver others into acting in its interest. There has been debate between this view and the view that signals are for conveying information, especially about the signaler's behavioral tendencies.

Marking behavior   Various conceptions of marking behavior are found in the ethological literature. Many writers restrict the term to olfactory marking, that is, to an animal's leaving chemical traces of urine, feces, or glandular secretions, which may continue to be effective after the animal has left the scene. Other writers use the term in a wider sense to include the use of acoustic and optical signals, and refer to the olfactory cases as scent marking.

All three forms of marking—chemical, acoustic, and optical—serve to advertise territory. Thus many frogs and birds and some mammals mark their territories by calling or singing. Visual marking is exemplified by the brightly colored males of the demoiselle dragonfly, which make conspicuous display flights around their territorial borders. Animals may also use olfactory marking to give aromatic distinctness to their own bodies or to label conspecifics such as the mate or offspring or fellow members of the social group to distinguish insiders from outsiders, as in social insects (see GROUP ODOR).

Markoff chain   See Stochastic analysis

Mate choice   See Sexual selection

Mate guarding   See Guarding behavior

Mating   See Copulation; Pairing

Mating march   Pairing promenade. A descriptive term for a form of pairing prelude in which the males and females run about for some time in tandem, one following the other closely or in direct bodily contact. The term is used mostly of ungulates, in which the males closely follow the females and move very stiffly (hence "march"). However, the term is also used of species in which there is indirect spermatophore transmission. Thus in scorpions and pseudoscorpions the male and female grip each other with their pincers and run back and forth until the male deposits his spermatophore and the female catches it up in her genital opening.

Mating system   Pairing system. The organization of reproductive relationships of a species or population. It includes the distribution of sexual attachments (see MONOGAMY; POLYGAMY), the kind of pair formation, the kind of pair bonding, if it exists, and the distribution of effort in parental care. In many species the mating system varies with environmental conditions. Thus African ostriches are sometimes monogamous and sometimes polygamous, depending on their age and breeding experience; in the North American caribou the nomadic populations of the tundra are monogamous, while the sedentary populations have harem formation and defense. The concept of mating system is subordinate to that of social system, but the two terms are not always sharply differentiated.

Maturation   Behavioral development that is independent of experience or practice. A behavior pattern is said to be a product of maturation if, in the course of its ontogeny, it becomes fully functional even when there is no opportunity for prior performance and hence acquisition of the skill through training or exercise. Maturation depends upon developmental processes in the central nervous system that are based on information encoded in the genome. Experiments to test for maturation involve preventing young animals from performing a particular activity at times when practice might be influencing its development, and comparing such animals with normally reared animals when they reach the age at which the behavior is completely developed. For example, the wings of young birds are restrained so that they cannot make the flapping movements that normally precede development of full flight. If little or no difference is found between the experimental and the control animals, the thesis of maturation is supported.

Critics of this kind of work have argued that the conclusion applies only to the experience withheld and leaves open other possible sources of experiential influence, some of which may be very difficult to rule out. On the other hand, if the experimental animals are retarded in their behavioral development, a proponent of maturation can argue that this does not rule out maturation, because the unnatural restraint could have had deleterious effects aside from preventing practice. If, during normal development, electrophysiological monitoring of the central nervous system reveals patterns of neural activity like those underlying a behavior pattern that the animal cannot yet perform because it is still lacking the bodily structures required, this too is claimed as evidence of maturational processes.

There is a possibility of confusion in the meaning of the word because of an earlier ethological usage. For example, Tinbergen (1951) used "maturation" to refer to the seasonal recurrence of behavior patterns having to do with reproduction (nest building, song), which, he said, must be considered separately from the initial growth of the underlying neural mechanisms. The word is now most often used for such growth processes, but a reader should keep in mind that in a particular context, maturation may be used in Tinbergen's sense.

Maze   Labyrinth. In ethology and experimental psychology a maze is a system of adjoining passageways used to investigate the learning capabilities of animals. One of the pathways goes from the starting point to a concealed goal box, where reward is offered. After some number of trials, an animal progresses from initially taking a more or less random sequence of right and left turns to following the "right" path. A maze's degree of difficulty amounts to the number of choice points it includes. The simplest is a T maze or Y maze, which requires only one decision. The usual measure of learning performance in maze experiments is the number of trials that an animal requires before it can find its way to the goal without entering any of the blind alleys. Another measure is the total number of successive turns an animal is capable of mastering in a complex maze.

The animals most often used in maze experiments are rats and mice, which live in a system of passageways under natural conditions, and so bring predisposing aptitudes to the experimental situation. Mazes have played a major part in research on learning, especially in the American school of behaviorism.

Meaning   See Communication theory; Message-meaning analysis

Mechanoreceptors   Sense cells that react to mechanical deformation, as from pressure, stretching, bending caused by the flow of air or water, vibration, or sound waves. Mechanoreceptors thus serve for the perception of tactile and auditory stimulation, linear and angular acceleration, gravitational direction, and postural positioning.

Melotope   A bioacoustical neologism for a habitat having specific characteristics that affect sound transmission within it. Quantitative comparisons have shown that an animal's sound production is often finely adapted to these characteristics. For example, its frequency range (wavelength) often conforms with that to which the habitat, because of its structure, is most permeable (the so-called acoustic window), and so conduces to the best possible sound transmission over distance. Beyond this, sound production is adapted to interfering noise in the environment. Thus reed warblers, which have to sing against a fairly constant background rustling of reeds, produce loud, harsh, rhythmic songs, while the inhabitants of closed woods, in which the general sound level is low, often have smooth, melodious, less vigorous songs.

Meme   A unit of human cultural transmission. Dawkins (1976) coined the term on the model of the gene, the unit of genetic transmission, and further developed the analogy between cultural evolution and organic evolution. Examples of memes are dress fashions, tunes, clichés, building styles, and ways of cooking. Like genes, they are replicated from one generation to another, can undergo sudden change resembling mutation, and may be subject to selection and undergo extinction. However, the mode of inheritance in cultural evolution (see TRADITION) is Lamarckian rather than Darwinian.

Memory   The capability of the central nervous system to store, in retrievable form, information derived from experience. The physiological processes underlying memory, and the physical nature of the memory trace, have yet to be fully elucidated. Brain research on various species has shown that specific memory functions can be associated with specific regions of the central nervous system, and also that the same memory content can be stored in several localities (see REDUNDANCY).

It is generally agreed that in higher vertebrates at least two forms of memory have to be distinguished, which are associated with different physiological processes: short-term memory stores information for periods as brief as a few seconds and at most a few hours; long-term memory can last for days, weeks, months, years, even the rest of a lifetime. Evidence of distinct memory processes comes from such findings as the effects of drugs on the different forms of memory (see PSYCHOPHARMACOLOGICAL DRUGS). Furthermore, information stored in short-term memory can be erased by loss of consciousness or strong electric shock, which long-term memories may survive. Long-term memory is believed to depend upon lasting chemical or structural changes at synapses or within neurons, effected during the short-term memory phase by neural transmission induced by the pattern of stimulus input. Recently the existence of an intermediate memory has also been considered.

Of the several conjectures about the nature of the memory trace (see ENGRAM), the most frequently entertained are that the information is coded in the form of neural circuits (see CELL ASSEMBLY) and that it is coded in the form of molecular structures. Both theories have their problems.

In biology the notion of memory is sometimes extended metaphorically from individuals to species. Thus "individual memory," acquired through individual experience, is compared to "species memory," which is information passed from one generation to another via hereditary factors. Such genetically transmitted information plays an especially prominent role in the behavior of lower animals (see INNATE; PROGRAM).

Menotaxis   See Taxis

Message-meaning analysis   In the study of animal communication, the working out of what a signal expresses about the signaler (the message) and what the signal, in context, conveys to a recipient (the meaning). (See also COMMUNICATION THEORY.) In practice messages are inferred from the situation and associated behavior of the signaler, and meanings are inferred from the response of the recipient. Smith (1965), who developed this approach in ethology, has argued that messages are mainly about the impending behavior of the signaler, each signal type encoding a single message or a very limited message set. Others have questioned whether signals always or generally convey information about behavior, arguing that some signals may be semantically versatile, encoding different content according to the context (see MANIPULATION).

Metabolism   A general term for all the chemical processes that keep an organism going, including those involved in digestion, respiration, and excretion. Metabolism comprises catabolic processes, which are constructive, as in protein synthesis, and anabolic processes, which are destructive, as in the breakdown of sugar molecules to make energy available when work has to be done.

Metamorphosis   See Larva

Microsomatic animals   Animals with poorly developed olfactory sensitivity. Among mammals, the term applies mainly to primates. Compare MACROSOMATIC ANIMALS.

Migration   Active travel by animals between locations relatively far apart. Passive transport by wind or current is excluded, even though it may cover a considerable distance. Migratory movements may be either two-way or one-way. Typical of the former are the seasonal migrations of animals that breed at high latitudes and winter closer to the equator, as exemplified by numerous species of birds, many whales, and monarch butterflies. Eels migrate from their birthplace in the sea to fresh water, where they spend most of their lives, then migrate back to their birthplace to breed and die. Salmon do the reverse. Spectacular examples of one-way migrations are the population explosions of locusts and lemmings, which serve for dispersal. The functions of migration are various, including exploitation of seasonal variation in food supply, adaptation to seasonal variation in climate, regulation of population density, and colonization of open habitat. The mechanisms governing migration include, for cyclical migrants, some form of endogenous periodicity in step with environmental change (for example, a circannual rhythm), and for dispersal migrants either a life history "program" (insects) or response to population density (mammals).

Some seasonal migrations are remarkable for the enormous distances traversed (Arctic terns breeding in the Arctic may winter as far south as the Antarctic, a round trip of at least 16,000 miles), and for their feats of navigation (golden plover breeding in Alaska fly 2,000 miles to Hawaii nonstop over open ocean). Some movements described as migrations are much smaller in scope, such as the daily vertical shifts of lacustrine plankton.

Migratory restlessness   Also referered to by the German term *Zugunruhe.* "The migratory activity of birds held in captivity, which, as a consequence of the thwarting limitations on flight, is expressed mainly as hopping, flapping and fluttering" (Berthold 1971). Migratory restlessness offers a good measure of readiness to migrate, so its measurement plays an important part in laboratory research on bird migration. Because migratory restlessness in a cage is undiminished even when all the normal releasing and orienting stimulation (such as changes in ambient temperature and/or photoperiod) is excluded, it presents a good example of an endogenously regulated process.

Mimesis   See Camouflage

Mimicry   A kind of imitation in which one kind of organism is conspicuously like another in appearance or behavior, to some biological end. Several types have been distinguished. The best known is Batesian mimicry (false warning coloration), in which the conspicuous advertisements of "protected" species (unpalatable or dangerous animals, or ones that can be caught only after excessive expenditure of energy) are adopted by unprotected species, as in the numerous cases of flies that resemble wasps or bees. As long as the mimic is sufficiently less numerous than its model (the species it imitates), it enjoys the protection from predator attention that the warning appearance provides for the model. In Mullerian mimicry, two or more unpalatable species converge in the way they advertise their nastiness. Looking alike, they do not have to be sampled separately for predators to learn to leave them alone, which reduces predator pressure below what it would be for each species. Peckhamian mimicry (aggressive mimicry) consists of imitations that predators present to lure prey, such as simulation of the food of the prey (for example, wriggling wormlike projections of the body, which angler fish and some species of snapping turtles use as bait to induce prey to approach within seizing distance) and counterfeit mating signals of the prey species: in some firefly species "femmes fatales" mimic the flash responses of females of a closely related species to the flashes of their mate-seeking males, and so decoy the males to their deaths.

Other examples of behavioral mimicry do not fit into any of the categories mentioned. A number of ground-nesting birds, such as plovers, perform a display called the "rodent run," in imitation of a running rodent, to draw the attention of an approaching predator away from the nest or brood (see DISTRACTION DISPLAY). More puzzling from a functional point of view are the instances of vocal imitation by birds such as parrots, mynas, and mockingbirds. The latter belong to a taxonomic family named after the habit of imitating the songs of other species: the mimic thrushes or Mimidae. See VOCAL MIMICRY.

Mixed motivation   A condition in which a behavior pattern is the expression of more than one behavioral tendency or motivation. This may be shown by the composite character of the behavior pattern. Thus many threat movements and postures consist of elements of attacking and escape actions, and the courtship behavior of many species contains attack and escape elements in addition to sexual components. The patterns attributed to a bonding drive may in fact be cases of mixed motivation.

Mixed singer   A songbird that, in addition to its species-typical song or songs, produces the songs or parts of songs of other species, usually close relatives, or combines them with its own in a medley.

Examples are the European goldcrests and firecrests. The term is also applied to dialect-forming birds that use, besides the population-specific dialect, song elements from another dialect. Mixed singers are not identical with vocal mimics. Mimicking species frequently include strange sounds, and, as a rule, the song elements come from many different species that need not be closely related to the mimic. Mixed singers use the songs of individuals of related species or populations and rarely adopt any foreign sounds.

As with vocal mimicry, no conclusive explanation has been offered for the existence of mixed singers. Interspecific territoriality is a possible function, at least for some species. Another suggestion is that the extraneous song elements are the result of erroneous imprinting during the time of song learning.

Mixed-species group A group composed of members of different species, usually ones that lead similar lives, such as grazing ungulates. Mixed groups are found mainly in mammals, birds, and fish. The biological benefit may be little different from that for groups composed entirely of members of one species. However, because of differences in sensory awareness from species to species, it is possible that mutual warning provides for a better defense against predators than occurs in homogeneous groups.

Mixed strategy See Strategy

Mobbing A general term for behavior patterns that many bird species direct toward a predator, such as a roosting falcon or owl. "While mobbing, birds of one or more species assemble around a stationary or moving predator (potentially dangerous animal), change locations frequently, perform (mostly) stereotyped wing and/or tail movements and emit loud calls usually with a broad frequency spectrum and transients" (Curio 1978). The function of mobbing is not yet clearly understood and probably varies. It serves in some way as a defense against predators, for example, by bewildering the predator (confusion effect: see FLOCK) and/or warning other individuals. Whether mobbing can induce a predator to leave the vicinity has yet to be demonstrated. Possibly it conveys to nearby conspecifics information about the danger posed by a predator.

Similar behavior, also sometimes described as mobbing, is known from many mammals, for example, baboons when they discover a predator (leopard or jackal) near the group, and American ground squirrels, which rush at snakes and even bite and fling sand at them.

Mocking   See vocal mimicry

Modal action pattern (MAP)   A recently introduced term for the smallest behavioral unit specifiable as a recurrent pattern of movement, that is, essentially what was previously referred to as a fixed action pattern. The term signifies that the movement sequences are differentiated by their normal morphology ("modal" here means concerning kind and manner), with no assumption of genetic fixity, which may falsely imply that the pattern is completely independent of environmental or peripheral influence. "(1) A Modal Action Pattern is a recognizable spatiotemporal pattern of movement that can therefore be named and characterized statistically. (2) It usually cannot be further subdivided into entirely independently occurring MAPs. (3) It is widely distributed in similar form throughout an interbreeding population" (Barlow 1977).

Modality   Sensory domain. A general term, stemming from human psychology, for stimuli perceived via a particular kind of sense receptor (light, sound, smell, taste, or touch), and therefore giving rise to sensations alike in quality to one another and distinct from all others.

Modification   Environmentally or experientially induced change occurring in the life of an individual organism. It thus contrasts with mutation in that it is not genetically determined. Modification may affect morphological characteristics (body size, shape, and color of animals, growth patterns of plants), physiological characteristics (temperature and light sensitivity), and behavioral characteristics, such as the manner of foraging, tool use, and vocal repertoire. Sometimes behavior modification can be transmitted from generation to generation through tradition.

Monogamy   Social organization in which males and females are paired one to one, at least for the duration of a reproductive cycle,

188

and for the purpose of reproduction. In contrast to polygamy, monogamy is almost always associated with parental care by both parents. It occurs in all vertebrate classes, and sporadically among invertebrates (desert woodlice, some crab species, some beetles). In mammals it is infrequent and has been established with certainty only in some carnivores (most canids, many mustelids), some rodents (agouti), some ungulates (dikdik, klipspringer), and some primates (gibbons, siamangs, titi monkeys). Of all vertebrates, birds are the most monogamous: more than 90 percent of all avian species have this mating system.

The duration of the pair bond varies from a single brood (as in many songbirds), to a whole breeding season or several breeding seasons (permanent pair bond). Lifelong fidelity has been reported in greylag geese and ravens. In many cases of monogamy, one of the partners usually gives way to the other (see RANK ORDER).

In most cases monogamy is associated with individual recognition between the partners. In some migratory birds, such as the white stork, swifts, and swallows, continual partnership is maintained through an attachment to the same place, to which the pair returns after separate residence in the wintering region. This state of affairs has been referred to as place fidelity (see FIDELITY TO PLACE) and anonymous monogamy.

Monomorphism   Uniformity of appearance and/or behavior within a species. Species are described as monomorphic when they exhibit no differences in appearance between the sexes (see SEXUAL DIMORPHISM), no division into castes, and no segregation of behavioral roles. Compare POLYMORPHISM.

Mood   See Motivation

Mood induction   See Social facilitation

Morphological support   In the evolution of display behavior (see RITUALIZATION), features of form and coloration have frequently developed concomitantly. Examples of such morphological support are the robin's red breast, the cockatoo's crest, the peacock's tail, the eyespots on the hind wings of many moths, and the swollen and brightly colored buttocks of sexually receptive female monkeys of many species. Usually specific postures or movements emphasize the presence of such features, and the combination functions as a releaser.

Mother-infant attachment   The especially strong bond between mother and young, which is prominent in mammals, most markedly in primates. Its evolutionary origins may be connected with the close bodily contact that the passively precocial primate young must maintain by holding on to the mother's hair (see CLINGING YOUNG). In many cases the bond persists through the period of dependence on the mother for nourishment. The close association between mother and young probably also serves, among other things, for the transmission of learned patterns. Disturbance of the mother-infant bond can cause deficits in behavioral development, which are categorized as features of the separation syndrome. The importance of the bond for later social development is most evident in the case of human children, as has been repeatedly demonstrated by human psychology.

Mother surrogate   See Surrogate

Motif song   See Song

Motility   A general term for an organism's capability of movement (with the exceptions of growth movements). It comprises locomotion, change of posture, and movement of parts relative to the body as a whole, including those of inner organs.

Motivating stimulus   See Priming stimulus

Motivation   Behavioral mood, drive; action readiness or tendency; the total complex of proximate causal factors governing what an animal is doing or about to do; compulsion or urge to behave in a certain way; control of behavior in animals, including people. As this by no means exhaustive list should convey, the meaning of this term ranges over a broad spectrum. It may be operationally specified in terms of behavioral probabilities, based on previous observation. But it may also connote what accounts for or underlies such probabilities, including external stimuli, internal stimuli, hormones, the state of central nervous arousal, and such variables as the time since last performance of the behavior in question.

It used to be assumed that the motivation of any behavior pattern could be expressed as a single value—its strength or intensity—at any point in time, equivalent to the ease with which the behavior could be elicited. This conception was supported by observations that the performance of consummatory acts (see END ACT) temporarily decreases the likelihood that the behavior leading up to it will be resumed, and also the releasability of the act itself; it was also observed that the tendency to resume the appetitive behavior and the releasability of the act increase as time passes, to the point that the threshold for stimulation may drop to virtual zero and the act occurs as a vacuum activity. However this conception of motivation as a unitary variable has problems associated with it, even when the variable is operationally defined in behavioral terms. Behavior of any sort can be quantified in a variety of ways; for example, one has to choose a time unit, for which there may be a considerable range of options. Research has shown that alternative measures of the "same" behavior often disagree, implying that different aspects of the behavior are at least to some extent determined independently of others. When the concept of motivation includes actual and hypothetical underlying causal factors and systems and forces, the assumption of unitary agency becomes even more difficult to sustain. Some behavioral scientists have gone even further and incorporated maturation processes, learning, and habituation phenomena, on the grounds that all such influences bear on present readiness to perform, but this is not accepted by many.

In spite of these problems of definition and inconsistency of usage, it is useful for ethologists and experimental psychologists to have a general term to cover the subject of proximate causal control of behavior, without commitment to a specific conception or theory. It is now common practice to talk of motivation in this way. Nevertheless, the problems covered by the term remain. See MOTIVATIONAL ANALYSIS; MOTIVATIONAL STATE).

Motivational analysis Investigation of the flux of a behavioral tendency as reflective of the causal state upon which it immediately depends. In ethology, as a rule, motivational state is not directly measurable, but it can be inferred from observation of the behavior (the magnitude and frequency of a specific activity) it is postulated to account for. For example, a change in motivational state is outwardly discernible when an animal responds differently to the same stimulus at different times, all other relevant external conditions being kept constant. Such study also involves analyzing how different activities vary in relationship to one another, for example, whether they are positively or negatively correlated with one another in timing or frequency of occurrence. The data are often quite complex and may require elaborate statistical computations to sort out, such as factory analysis and stochastic analysis. Even then, a priori assumptions have to be made, which leave inferences of underlying motivational organization open to question.

Motivational change   In an evolutionary sense, a change in the underlying causal control of a behavior pattern in accordance with a change of function. Thus the courtship feeding of many birds corresponds in form to the manner in which parents feed their chicks. However, it has taken on a new function and consequently is no longer governed by the motivation underlying parental behavior, but by sexual motivation. The term is virtually synonymous with emancipation, except that the latter is usually restricted to causal change concomitant with ritualization and hence the evolutionary emergence of communication displays.

More commonly, at least in ethological writing in English, motivational change refers to a short-term shift in motivational state, as when fear-eliciting stimuli overrule factors causing grooming behavior.

Motivational energy   Lorenz (1950), observing such things as the refractoriness immediately following a consummatory act (see END ACT), threshold lowering with passage of time, and vacuum activities, postulated that instinctive activities are impelled by spontaneous generation of "action-specific energy," which is used up in performance of consummatory acts and accumulates between their performance. He used a hydraulic analogy to represent his conception of the mechanism. Tinbergen (1951) elaborated this conception into his well-known hierarchical theory of instinct, in which the behavioral energy was referred to as "motivational impulses" and which was also pictured as working in accordance with quasi-hydraulic principles. These conceptions of motivational energy, which bear some resemblance to Freud's concept of psychic energy, have become less plausible following criticism from ethologists and others. Most ethologists have given up thinking in such terms.

However, recently the notion of motivational energy has been revived in a different conceptual framework. McFarland (1971) has presented a concept of motivational systems that conforms to the mathematical expressions describing mechanical, electrical, hydraulic, and other physical systems, in which energy is rigorously defined as the capacity for doing work. The parallels have enabled him to formulate correspondingly rigorous definitions of motivational energy, motivational state, and other variables, parameters, and functions, and so to use the mathematics of such powerfully integrative fields as statistical mechanics to bring order to and generate predictions in the study of the control of behavior. As McFarland admits, the value of this approach has yet to be determined, and there are problems about the descriptions of behavior it assumes. Nevertheless the application of control theory to behavior may have given the concept of motivational energy a new lease on life.

Motivational impulse   See Motivational energy

Motivational state   The combination of physiological and perceptual factors underlying an animal's behavior or impending behavior at a particular time. A recent conception pictures motivational state as located in a multidimensional space, the axes of which are formed by the relevant physiological and stimulus variables, whose values are affected by the consequences of an animal's activity. This way of representing and computing motivational state avoids some of the problems that confounded unitary conceptions of drive and motivation.

Motivational system   The causal regulation underlying a particular kind of behavior. Older conceptions, such as the instinct theories of classical ethology and unitary drive theories (see MOTIVATION) supposed a separate regulatory system for each of the major functional categories of behavior. (See FUNCTIONAL CYCLE.) These theories are now generally considered not to have dealt with the full complexity of behavioral control, in particular the extent to which interactions among different physiological processes may be enmeshed in the causation of behavior. More recent approaches have drawn on the control systems theory developed by engineers for dealing with complex machinery. As a consequence the concept of motivational state has superseded the notion of a specific drive for each kind of behavior.

Motor deprivation   See Isolation experiment

Motor nerves   See Efference

Motor pattern   Movement pattern. Behavior described in terms of spatiotemporal change in body configuration, for example, foot flexion, fin fanning. (Compare FIXED ACTION PATTERN; MODAL ACTION PATTERN.) Another, contrasting way of specifying categories of behavior is in terms of functions or consequences, such as retrieving, nest building, or prey catching. Some terms (head scratching) connote both the motor pattern and the function it effects. However, the two ways of describing may map differently onto the same behavior, one making distinctions where the other would not; for example, the motor pattern of running may be an instance of fleeing or an instance of chasing; and the action of fleeing may be an instance of running or an instance of swimming. The distinction is important where there is the possibility of confusing questions of causal control, pertaining to motor patterns, and questions of ends served, pertaining to functionally distinguished actions.

The Eshkol-Wachmann movement notation, originally devised for ballet, has brought refinement and precision to the description of animal motor patterns, including complex stylized interactions between two animals.

Mount bouts   See Mounting

Mounting   In mammals, the male's adoption of the position for copulation, which in nearly all cases consists in standing on the hindlegs and grasping the flanks of the female from behind. Several mounts may be required, with or without pelvic thrusting and intromission, before ejaculation can occur. The composition and duration of such mount bouts vary between species in ways that have adaptive correlates. In some species, especially among primates, mounting also serves a more social function: high-ranking individuals display dominance by mounting subordinates, either of the same or of the opposite sex (see PSEUDOCOPULATION).

Mouth breeder   A kind of fish that carries its eggs and young in its widely distended mouth cavity and so protects them from predators. During this time the young live on their yolk supply and leave their protective shelter only when that supply is exhausted; in some species they continue to return for some time afterward, especially in times of danger. Mouth breeders produce fewer but larger eggs than substrate-breeding species, which lay their eggs on the bottom. The mouth breeders are mostly cichlids or labyrinthine fishes (anabantids). This kind of parental care has evolved independently in these two groups and in some others, and even within each group has repeatedly arisen independently in quite distantly related species. Its distribution is thus a consequence of convergent evolution.

Movement   In ethology this term can mean either motor pattern or locomotion. It can also refer to change of location, as in "migratory movement."

Mullerian mimicry   See Mimicry

Multimale group   A term applied to primates, referring to a mating system in which, in contrast to single-male groups, at any given time several adult, sexually mature males live together with the females and all stages of young animals. Such social organization is exemplified by many baboons and macaques. See also HAREM.

Multiple choice test   See Choice test

Multipurpose movement   Lorenz (1937) introduced the term *Mehrzweckbewegung* for behavior patterns (as a rule simple and brief) that appear in more than one functional system. Foremost are the motor patterns of locomotion, which occur in connection with foraging, territorial defense, migration, indeed in almost all kinds of activity. When an animal goes from one such activity to another via

195

a multipurpose movement common to both, the movement may be referred to as a transitional action.

Mutation Change in the genetic makeup of a cell. There are two types: chromosome mutation, in which chromosomes are multiplied or rearranged through breakage, inversion translocation, or some other kind of shuffling; and gene mutation, in which the molecular structure of the DNA is altered, specifically the selection or ordering of bases on the DNA strand. Under natural conditions mutations occur randomly and spontaneously, that is to say, without our being able to tell what causes them. However, their incidence can be experimentally boosted, for example by subjecting cells to short-wave radiation and other mutagenic agents. Organisms carrying mutations are referred to as mutants. Numerous mutant strains have been selectively bred in genetic research on such animals as fruitflies and mice. The mutant traits provide genetic markers enabling geneticists to make genotypic identifications in breeding or selection experiments. Although the majority of mutations are deleterious, mutation furnishes raw material upon which selection can act, and so is an important factor in evolution.

Mutualism See Symbiosis

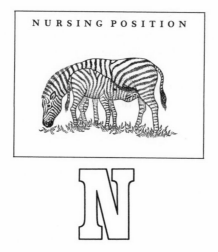

N

Natural selection   See Selection

Nature-nurture issue   See Innate

Navigation   Determining the course to take in long-distance travel, as in migration and homing. It involves assessment of position and goal direction, perhaps in unfamiliar terrain. The study of animal navigation has revealed that some animals possess internal clocks and are able to use the sun, star patterns, and a variety of other geographically regular features as navigation guides. See ORIENTA-TION.

Neck bite   A bite directed at the neck of another animal. Many felid carnivores use it to kill their prey, in contrast to other carnivores that shake the prey to death. A gentler form of neck bite occurs in the retrieving of young by various rodents and felid carnivores and in the copulation of many carnivores and birds, in which the male grips the female with its teeth or beak (pairing bite).

Necrophoric behavior   The carrying of corpses of dead conspecifics. Ants carry dead nest mates out of the nest, and parent birds remove dead nestlings from the nest. Cases have also been reported of elephants and of primate mothers carrying the bodies of dead infants about for days after their demise, as though in the hope of their recovery. Elephants are noted for carrying and "playing" with bones of their dead and for an apparent fascination with decaying carcasses.

Negative conditioning   Negative reinforcement. Use of aversive stimulation, such as electric shock, to change an animal's behavior. There is some inconsistency about what the reinforcement consists in. In some contexts the punishment itself seems to be regarded as the reinforcer, by associating pain with a situation or response. However, when experimental psychologists talk of negative conditioning or negative reinforcement, they usually mean that relief from or avoidance of pain is the reinforcer, conditioning the preceding behavior to the situation. The behavior may be either passive avoidance—staying clear of whatever has caused distress—or active avoidance—performing some response to prevent or reduce distress, as when a rat learns to press a lever in response to a stimulus signaling impending shock. In natural circumstances the experience of illness caused by poisonous food can lead to taste aversion (bait shyness).

Negative feedback   See Feedback

Neophobia   Fear of novelty. The tendency of an animal to steer clear of any new object or situation it encounters. In mammals especially, neophobia is often in contention with curiosity (see EX-PLORATORY BEHAVIOR), and an animal confronting novelty will alternately approach it and withdraw from it and appear to be in a state of heightened alertness and arousal. As with the orienting reflex, fear of an object usually wanes as it ceases to be strange if nothing of consequence is associated with it.

Neoteny   In biology, a general developmental slowing down during evolution, such that an animal becomes sexually mature before it is physically fully grown. The same consequence can result from a speeding up of sexual maturation relative to physical maturation, a process known as progenesis. The evolutionary outcome—ancestral juvenile characteristics in adult stages of descendants—is known as paedomorphosis.

A developmental retardation comparable to neoteny, and having direct significance for behavior study, has been observed in some higher mammals and birds. It consists of a trend toward a longer or slower juvenile period of development. The reasons for the trend are not fully understood. One possibility is that the lengthened development time is necessary for accumulation of metabolic reserves sufficient for the transition to independence (see WEANING). More likely its main significance is in the extension of time available for gathering experience through play, exploratory behavior, and the transmission of tradition. In species where the young remain for lengthy periods with their parents or the social group, it is important that they be easily recognizable as juveniles (see JUVENILE CHARACTERISTICS).

Nepotism   See Altruism

Nerve impulse   Action potential. The firing of a neuron. It consists of a transient localized change of electrical state of the nerve cell membrane, in which the inside of the cell goes from negative to positive and back again with respect to the outside; the same change is induced in neighboring membrane and propagated from point to pont, typically from the cell body along the axon to its synaptic terminals. Nerve impulse transmission is the mode of communication in the nervous system, information being coded in such variables as firing rate, length of volleys, and neurons activated.

Nest building   Many animals construct receptacles or shelters within which they lay and incubate eggs or spend significant parts of their lives. Most birds build nests for incubation and brooding, ranging from saucer-shaped scrapes in the ground or sand to elaborately woven structures requiring complex sequences of finely adjusted movements. Hole-nesting birds seek out or dig cavities in trees or banks, within which they may construct a cup of softer material. The cave swiftlets (*Collocalia*) use saliva in the construction of their nests, some of which have so little other material as to be edible. A few species build communal nests, the most elaborate being those of the social weaver (*Piletairus socius*) of South West Africa.

Some mammals also make nests to hold the young or for their own comfort, but they are usually crude accumulations of material compared with a typical bird's nest. The nest may be situated within a burrow. A number of fish species construct nests. The male three-spined stickleback digs a trench, roofs it with plant material, and contrives to get a female to lay eggs in it; labyrinthine fishes, such as gouramis, make bubble nests, which hold the eggs and very young fry. Among invertebrates the master nest builders are the social insects, which present a great variety of structures, some, such as ant hills, of considerable size. In some species the insects live all or a large part of their lives in the galleries of their constructed environment.

Nest hygiene Nest-cleaning behavior. It is especially pronounced in birds, such as songbirds, in which the young remain in the nest for some time after hatching (see NIDICOLOUS NESTLINGS). Typically such nestings defecate only when a parent is present, as just after feeding. The fecal matter is enclosed in a sac, which the parent picks up as soon as it emerges from the chick's cloaca, and flies off with it, dropping it some distance from the nest. In many species the cloaca is surrounded by a conspicuous ring of white feathers, which, when the nestling lifts its rump, acts as a releaser causing the parent to prod the chick, which releases defecation by the chick.

Nest hygiene prevents fouling and caking of the nest and nestlings, prevents or reduces the infestation of parasites, and prevents the conspicuousness that would result from accumulation of white fecal guano around the nest. Many species of birds also carry egg shell remains from the nest, especially where their conspicuousness might draw attention to the nest.

Nest hygiene is also part of the behavior of social insects and includes removal of debris and the remains of dead individuals from the nest.

Nesting association There are numerous instances of birds nesting close to other species, and also of other species living in birds' nests. These associations, most of which are symbiotic, take numerous forms. Mixed colonies of two or more species of birds are presumably mutually beneficial because of combined predator defense. Some birds nest in the nests of termites or ants, presumably for the shelter of the structure, with no benefit to the insects. Others nest close enough to ants, wasps, or bees to share the protection of the aggressiveness of these insects toward intruders. A more balanced symbiosis exists where insects living in birds' nests help to keep the nest clean in return for the shelter it affords them. In New Zealand some shearwaters (*Puffinus carneipes* and *P. bulleri*) share their burrows with tuataras (*Sphenodon punctatus*). Some birds have established nesting associations with humans, such as swallows and storks that use buildings for nesting sites.

Nesting symbol   A piece or bunch of nesting material (grass blades, twigs, straws, leaves, or pebbles), which the male in many species of birds presents or shows to the female, often with accompanying bill or head movements, prior to or during mating behavior. This behavior is part of courtship and probably serves to suppress attack and fleeing tendencies. In some species, such as the crested grebe, both partners hold nest material in their bills during courtship ceremonies.

Nest relief   The taking over of the incubation of eggs or brooding of chicks by the partner of a pair of birds. In species in which both sexes share in parental activities, the male and female relieve each other alternately at the nest. The schedule of relief is species-specific within certain limits. Thus many small birds relieve one another at intervals of only a few minutes; pigeons and doves, in contrast, do so only twice a day, in the morning and the evening. Many species have special behavior patterns called relief ceremonies, which, among other things, inhibit or divert aggressive tendencies and may be regarded as appeasement behavior.

Nest relief also occurs among monogamous fishes, in which both partners of a pair contribute to the parental care. Here too the interactions usually involve a highly ritualized ceremony, but these interactions are less well understood than the nest relief of birds.

Neural inhibition   See Inhibition

Neuroembryology   See Behavioral embryology

Neuroendocrine cell   See Neurosecretion

Neuroendocrinology   The study of hormone secretion by cells in the central nervous system and its function in the regulation of metabolism, growth, and behavior. See NEUROSECRETION.

Neuroethology   A part of behavior study that is concerned with the processes in sense organs and the central nervous system that underly performance and control of behavior patterns. Questions dealt with in neuroethology include how the central nervous system patterns and regulates movement; which neurophysiological processes affect the motivational states of animals; how information derived from experience is stored in the brain; (see ENGRAM; MEMORY); and how the particular key stimuli are selected from the welter of stimulation impinging on an animal from its environment (see STIMULUS FILTERING).

Neurohormone (neurohumor)   See Neurosecretion

Neuron   Nerve cell. The basic unit of the central nervous system. A cell that can generate, propagate, and transmit electrical signals and thereby communicate information within the body. See ACTION POTENTIAL; NERVE IMPULSE.

Neurosecretion   Production of physiologically regulative substances by cells in the central nervous system. As it stands, this definition could cover nearly all CNS neurons, since chemical discharge is the rule in synaptic transmission. This may be justifiable, because the difference between neurons and neurosecretory cells is one of degree rather than kind. However, neurosecretion connotes the production of chemicals that have their effects some distance from the site of origin; for example, in vertebrates the neurosecretory cells in the hypothalamus have their target cells in the anterior pituitary. When the neurosecretory cells deliver their secretions into the bloodstream they are functioning as endocrine glands, and so are referred to as neuroendocrine cells, and their secretions as neurohormones (also neurohumors). In many invertebrate species (insects, crustaceans, cephalopods), neurohormones contribute to regulation of such processes as molting, metamorphosis, circadian rhythms, and color change.

Neutral object   A "substitute" object to which action is directed in conflict situations (see REDIRECTION). However, the term "substitute object" also refers to what may elicit action as a consequence of the lowering of a threshold, as when a rat deprived of nest material treats its own tail as nest material. The two situations differ in a way that requires drawing a distinction: in the case of threshold lowering the substitute takes the place of the "right" object because the latter is unavailable; with redirected movement the relevant object is present but intimidating, so action toward it is blocked and diverted elsewhere.

Niche   See Ecological niche

Nidicolous nestlings   In birds, chicks that remain in the nest for some time after hatching, where they receive food and warmth from their parents. They are therefore altricial young. In many cases they are hatched naked, with their eyes still sealed, and are unable to stand and walk. Songbirds provide typical examples. In gulls, however, the chicks are covered with down feathers at hatching, have open eyes, and can stand and walk. Nevertheless they remain in the nest for at least a few days (in the case of kittiwakes they stay until fledging) and are fed and brooded by their parents. Such cases have been described as semiprecocial. Compare NIDIFUGOUS NESTLINGS; PRECOCIAL ANIMALS.

Nidifugous nestlings   Young of precocial species of birds. The root of the term means "leaving the nest." When hatched, the chicks are at a relatively advanced stage of development and can actively follow their parents after only a short time. They are capable of feeding themselves and do not require their parents to do more than lead them to food. Waders are good examples of nidifugous birds. The contrasting category is nidicolous nestlings.

Nonsocial   See Solitary

Nonverbal communication The use by people of nonlinguistic means to convey signals to one another. Gesture, facial expression, posture, and movement can indicate mood, serve as greeting, signal assent or dissent, and mean a variety of other things. Crying and laughing, shouting and crooning, and numerous other sounds can also communicate without words. And touching conveys messages. Nonverbal signals may be deliberate, as in waving goodbye to someone, or involuntary, as in blushing. Some nonverbal signs are universal human expressions, and these may have homologues in other primates. Examples are smiling and "eyebrow flashing". Other forms of nonverbal communication vary with the culture and may be highly conventional. The study of human body language—the patterns of gesture, movement, posture, and orientation in social interaction—is sometimes referred to as kinesics. True nonverbal communication does not include such systems as gestural sign languages or the whistling languages that have been independently developed by numerous peoples, for these systems are coded to a verbal base.

Number concept See Counting ability

Nuptial dress See Advertising dress

Nuptial flight The communal flight of the male and female in some species of insects for the purpose of mating. Copulation occurs either during flight or upon landing. Especially impressive are the nuptial flights of social insects in which many hundreds or even thousands of reproductively primed animals sometimes swarm in the air together. The flights that serve to initiate pairing in many species of birds are referred to as courtship flights.

Nursing position Suckling posture. The body position adopted by young mammals during suckling. It is adjusted to the mother's body position during nursing within certain limits that are species-characteristic. In the majority of ungulates the calves stand under or alongside the mother and face in the opposite direction from her. In most rodents and carnivores the young suckle lying down. Young primates and bats cling firmly to the mother during nursing (see CLINGING YOUNG). In some cases, the nursing position differs between closely related species. For example Old World swine suckle lying down, while the New World peccaries suckle standing up.

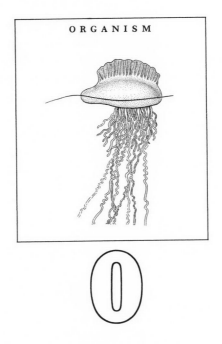

O

Object imprinting   See Imprinting

Ontogenesis   Ontogeny. Individual development. Development of
an organism from fertilized egg to mature adult. Some writes consider
ontogenesis to continue on to senescence and death. It thus includes
the prenatal embryonic and fetal development in the egg or in the
mother's body, the juvenile stages between hatching or birth and
the sexually mature, fully grown adult, and the subsequent processes
of deterioration with age (senility). The development of behavior is
referred to as behavioral ontogeny. Ethological study focuses especially on development from birth to sexual maturity, for investigation
of the many changes that occur during this period can contribute to
understanding of learning and maturation processes and the interplay of genetic and environmental factors in behavioral development.
Study of prenatal development has revealed that learning can occur
prenatally (see BEHAVIORAL EMBRYOLOGY). Developmental study
can also contribute to understanding of the phylogenetic origins of
behavior, for transient feaures in the behavior of a young animal are
sometimes the same as or similar to behavior in the repertoire of a

related species presumed closer to the common ancestor (see RE-CAPITULATION).

Open-field test   A method frequently used for quantifying various aspects of behavior in the laboratory, especially of rats and other rodents, and occasionally primates and birds. The open field is a space to which the experimental animal has not been previously introduced, usually the size of the animal's living cage, square or rectangular in shape, surrounded by walls, and divided by lines into equal-sized quadrants. Examples of variables measured in an open field test are the animal's locomotor activity in relation to previous environmental conditions or experience, scored as the number of quadrants entered per unit time, and the animal's "general emotionality" or level of anxiety, as indicated, for example, by the amount of urination or defecation.

The interpretation of such experiments requires care, for in some cases the animal's reactions in such an artificial situation have little correspondence with the behavior of the species under natural conditions.

Operant conditioning   See Conditioning

Operational definition   A meaning that is restricted to observational or measurement or manipulation procedures. The term was first used in physics, but it has also been applied in the study of behavior. Thus "behavioral tendency" might be defined operationally in terms of the level of stimulation necessary or sufficient to elicit the behavior, with no "surplus meaning" implied about unobservable internal states or processes. Strict operationism would bar the use of hypothetical constructs. Behaviorists insisted on operational definitions of psychological terms, although they were not always consistent in their adherence to it. In ethology the replacing of the term "action-specific energy" with "specific action potential" was an example of operationalism.

Optimization theory   Optimality theory. In behavioral ecology, an approach to the question of whether and how organisms presented with alternative strategies for exploiting their environments select the option that maximizes benefits and minimizes costs. The theory has been directed mainly at optimal foraging patterns.

Organ   Body part. A functionally distinct structure composed of tissues and constituting part of a physiological system, for example, heart, liver, and brain. For control of behavior, the most important organs are the sense organs, the central nervous system, and some of the endocrine glands.

Organism   Individual living thing. A multicellular organism is a harmonious integration of cells, tissues, organs, and systems. In certain animal groups, however, the degrees of dependence and independence of parts make it difficult to decide where the line should be drawn between individual organisms and assemblages. For example, the Portuguese man-of-war (*Physalia*) is regarded as a colony of coelenterate "persons," yet it looks and behaves so much as a unit that it is difficult to resist the impression that the whole is the individual organism. Sponges present a contrasting case of cells so loosely bound together physiologically that they might be considered collectives rather than individuals. In some insect societies the degrees of dependence among the members suggest analogy with the cells of a body: hence the concept of the super organism or supraorganism.

Orians effect   See Communal courtship; Fraser Darling effect

Orientation   The ability of organisms to direct their body position and locomotion with respect to the locations of objects and forces in the environment. Orientation depends upon sensory capacity, and different kinds of orientation are distinguished according to the sensory modality involved. An oriented movement is called a taxis, so we have, for example, phototaxis, thigmotaxis, galvanotaxis, and geotaxis, according to whether the orientation is based on light, tactile stimulation, electrical field forces, or gravity. Another difference is the direction in which an animal heads in relation to the stimulation. If an animal heads toward the stimulus or ascends the stimulus gradient, it is showing a positive taxis; if it heads away from the stimulus or goes down the stimulus gradient it is showing a negative taxis. A third possibility is menotaxis—heading at an angle to the stimulus source. Further distinctions are drawn according to the animal's means of registering and reacting to stimulation. In kleinotaxis the organism has only one sensor at its disposal, which it aims in different directions successively, adjusting its heading according to differences of intensity registered from moment to moment. The flagellate *Euglena* uses its single eye spot in this way by swimming in a spiral path and sampling the light intensity on all sides as it goes. People have to resort to a form of kleinotaxis when they try to locate something by smell. Tropotaxis is orientation based on registration of stimulus intensities at paired, symmetrically situated receptors. If the simultaneous comparison discloses an imbalance, the animal adjusts its position until inputs from the two sensors are equal. Planarians such as *Dugesia* have bilaterally and symmetrically placed eye spots at the anterior end that enable them to orient with tropophototaxis. People can tell the direction of sound tropotactically. Telotaxis, which depends upon perception of objects at a distance, is possible only for animals with sense organs capable of registering forms in space. It has evolved in only three groups in the animal kingdom—arthropods, cephalopods, and vertebrates—and in most cases only in the visual modality. However, bats and a few other vertebrates use echolocation, some snakes have infrared eyes, and some fish can sense deformations of self-generated electric fields; the resulting patterned perception enables telotaxis comparable to that of patterned vision.

Orientation is, of course, essential to navigation. Among the more sophisticated forms to be found in animals is sun-compass orientation; the animal uses the position of the sun in the sky as a reference for direction finding, compensating for the sun's travel by changing the angle between the path taken and the direction to the sun in accordance with time as told by an internal clock. Honeybees and diurnally migrating birds are the best-known examples of animals that can do this. Only a little less remarkable are the nocturnal migrants that can use patterns of fixed stars as navigation beacons. There is now evidence that birds and other animals use alternatives to vision for orientation during navigation; among the more likely possibilities is sensitivity to the earth's magnetic field.

Orienting reflex   Orienting response, startle response. Reaction to a novel stimulus, consisting of attentive scrutiny of it (see ATTEN-

TION), increase in arousal, and readiness to take flight or defensive action. The response wanes if nothing of significance proves to be contingent on the stimulus (see HABITUATION). The term derives from the Russian scientist E. N. Sokolov.

Orthokinesis   See Kinesis

Ortstreue   See Fidelity to place

Ova   Female gametes; egg cells produced in the ovaries of female animals.

Ovary   Female reproductive organ. In addition to producing ova, it may serve an endocrine function (see HORMONE).

Ovulation   Follicle rupture, egg release. The release of ripe ova from their enveloping capsule in the ovary. This covering is referred to as the egg follicle (in mammals, the Graafian follicle). The follicle provides the ripening egg with nourishment and also secretes estrogen, and therefore has an influence on behavior. In many mammals an influence in the reverse direction has also been recognized, in consequence of which mammals are categorized as spontaneous, or automatic, ovulators and induced ovulators. Spontaneous ovulation occurs in the majority of species, including primates, and hence human beings. The follicles rupture spontaneously as part of a species-characteristic cycle of ovarian activity. The length of the cycle is about twenty-eight days in humans, four days in rats, four months in domestic cats. Induced ovulators include shrews, rabbits, ground squirrels, mink, and ferrets. The release of the egg is not automatic but occurs some time after successful copulation—some ten hours in the case of rabbits. Here is an instance where the occurrence of a behavior pattern—mating—is a prerequisite for a physiological process—ovulation. The control of ovulation is effected through female sex hormones, in particular the gonadotropic hormones of the pituitary, estrogen and progesterone.

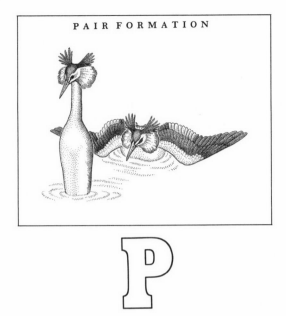

PAIR FORMATION

P

Pack   See Group formation

Paedomorphosis   See Neoteny

Pain-induced aggression   See Aversion-induced aggression

Pair bond, pair bonding   The tie between opposite-sex partners, manifested in their living together continuously for some length of time, in contrast to transitory liaisons, which last only until copulation is completed. A pair bond may last just to the end of a breeding period, that is, until the young depart to their independent lives, or it may persist beyond that time. Seasonal pair bonding allows for the sharing of parental care by both parents and for defense of the territory or protection of the brood (for example, through distraction display) by the males. For the persistence of pair bonds beyond the breeding period there is so far no single explanation. Among the advantages mentioned in the literature are the following: males can guard their females from the attention of other males; the partners jointly may be more successful than either alone, as in competition with others for food or assisting each other in aggressive interactions; and lastly it can obviate the necessity for new pair formations at the beginning of each reproductive season, thus promoting quicker synchronization of the pair's reproductive states and so getting the advantages of an early start in, for example, competition for territory, nest sites, and brooding places. In the cliff-nesting kittiwake (*Rissa tridactyla*), for example, previously formed pairs lay eggs earlier in the spring and have greater breeding success than newly formed pairs.

In most cases the sexual bond is between one male and one female exclusively, as the term "pair bonding" implies; that is the social system is one of monogamy. But in many species lasting heterosexual bonds are formed involving a larger number of individuals, as in the social structure referred to as a harem (see POLYGAMY).

As a rule species with pair bonding have special behavior patterns, called bonding behavior, for reinforcing the cohesion of the pair. In some cases there appears to be a special bonding drive.

Pair formation   Establishment of a pair bond. The behavior patterns involved are mainly those of courtship and mating, but there may be contributions from other categories, such as threat behavior. In most species the male initiates pair formation, but the female often makes the final decision on whether or not a bond is formed. In only a few species, such as the button quail, painted snipe, and phalaropes, does the female take the initiative in pair formation. In some species, especially those with very long-lasting pair bonds (see MONOGAMY), as in many ducks and geese, pair formation may occur some weeks or even months before the first copulation.

Pairing   In some contexts, synonymous with copulation or sexual coupling; in other contexts, pair formation or forming of a pair bond. "Mating" has similar variations in meaning. In both cases the context usually conveys which meaning is intended.

Pairing behavior   See Copulation; Courtship

Pairing territory   In some species of birds, including some gulls, pair formation takes place away from the site of the nest and the rest of the reproductive activities. Such a preliminary defended area is called a pairing territory. In the black-headed gull (*Larus ridibundus*), pairing territories are set up by birds starting courtship before the gullery site is occupied; once the birds settle in the gullery the pairing and nesting sites are the same in most cases. More distinct types of pairing territory are the leks established by some insects (Odonata), birds (ruffs, black grouse), and mammals (Uganda kob).

Paleoethology   A little-used term for the study of the behavior of extinct animal species (see FOSSILS).

Paradoxical sleep   Rapid eye movement sleep (REM sleep). Occasions during sleep when the arousal level, as indicated by an electroencephalogram, is elevated and the eyes move rapidly. Human subjects wakened during such episodes report dreams, but do not when they are wakened at other times. Paradoxical sleep has been detected in a number of mammal species.

Parallel evolution   Evolutionary change that follows more or less the same course in species that derive from common ancestors but are reproductively isolated from one another. Parallel evolution is a consequence of selection acting similarly but independently on separate lineages with a common evolutionary origin. It contrasts with convergent evolution (see CONVERGENCE), which gives rise to similar adaptations in unrelated organisms, and adaptive radiation, which is divergent evolution in descendants of a common ancestor.

Parasitism   A relationship between two organisms in which one, the parasite, finds its food and reproductive conditions inside (endoparasite) or on (ectoparasite) the body of the host, which consequently often suffers injury or debilitation. Parasitism usually involves a complex of special adaptations in body structure, physiology, reproductive biology, life history, and behavior. An example of behavioral adaptation is the food begging of many beetle larvae, mimicking the behavior of the ants they exploit; as a result they are tolerated in the ant society and even provided with food.

Special forms of parasitism are brood parasitism and social parasitism.

Parental care Protecting, nourishing, and nurturing of young. Parental care includes supplying food; attending to the state of skin, feathers or hair; removing feces and other waste (see NEST HYGIENE); camouflaging, such as covering eggs by ground-nesting birds; transporting the young to places of safety (see RETRIEVING); and guarding, defending, and leading the young. In egg-laying animals it also includes incubation of the eggs and, in many species of fish, the active aeration of the eggs by "fanning." Besides its immediate contributions to the welfare of the young, parental care also provides the possibility of learning from adults, especially when there is a long period of dependence (see TRADITION). As a rule the term applies only to direct contact between adult and young (compare BROOD PROVISIONING).

Parental care occurs in all classes of vertebrates and in some invertebrates, most commonly in arthropods (insects, spiders, myriapods, crustaceans) and cephalopods. In most of these species parental care is carried out by females. However, in many species (most commonly those that are monogamous) both sexes participate, and in some cases the male takes the greater share or even does it all (stickleback fishes). Parental care by males is referred to as paternal care and, especially in the primate literature, as male care if the care giver may be an individual other than the father. When both adults participate in parental care there may be some division of labor.

Parental investment "Any investment by the parent in an individual offspring that increases the offspring's chance of surviving (and hence reproductive success) at the cost of the parent's ability to invest in other offspring" (Trivers 1972). In addition to the physiological stress, parental investment includes the costs in time and energy and risk spent in brood provisioning and parental care. In the majority of cases where there is a difference, the female parent has a greater investment in her offspring than the male parent; as a rule, more of her substance goes into the manufacture of eggs, especially if food reserves such as yolk are included, compared with what the male puts into sperm. Also, in species with parental care, more of the female's time and energy than the male's is used, typically. As a rule a male can inseminate females without appreciably lessening his chances of doing so again, but when a female is inseminated the consequences for her subsequent reproductive life may be considerable. An extreme imbalance between the sexes exists in mammals, where the females, in addition to bearing the young, are burdened with having to nurse them on milk, which they alone produce. In species in which the females are capable of breeding a number of times, parental investment values suggest that at the beginning of her reproductive life a female should abandon her brood and save herself if confronted with the choice, but in the same situation at the end of her fertility, should sacrifice herself for the sake of her brood. Differences in parental investment can thus account, in evolutionary terms, for some of the differences between social systems, and for differences of strategy within a social system, such as frequently obtain between males and females.

The concept of parental investment originated in the literature of sociobiology in discussions of inclusive fitness and kin selection. From this came the notion that conflict between the demands of the young and the cost to the mother of continued parental care might lead to conflict between offspring and parent toward the end of postnatal dependence (see PARENT-OFFSPRING CONFLICT; WEANING). The metaphors of sociobiology, such as investment, cost, benefit, and tradeoff, have revived the analogy drawn by Darwin between evolutionary biology and economics.

Parent-offspring conflict A sociobiological notion related to kin selection theory which holds that there may be a conflict of interest between a young animal and its mother or parents (see FITNESS). According to this the "objective" of the young should be to get for itself the greatest possible parental investment from the mother (or parents); at the same time the parent must restrict such investment in order to maximize further reproductive output. The degree of the conflict should be different for relatively young parents, whose reproductive potential is high, compared with parents approaching the end of reproductive life, who can afford to invest more in young that may be their last. Also the age of the young is a factor; the older they are, the more they can look after themselves and the less they benefit from parental care; conversely, for the mother, demand increases as the young grow.

Numerous observations are consistent with this thinking. Thus the parental care of cats and dogs can be divided into three phases: during the first phase the mother takes the initiative in making contact with the young and succoring them; in the second phase the initiative comes from both mother and young; and in the third and final phase, shortly before weaning, the young try to get more nurturing than the mother, who sometimes reacts aggressively or evasively, is ready to give. A similar development, which can finally merge into a weaning conflict, has been described for rats, a large number of primates, and various species of birds.

Partial reinforcement Rewarding an animal for only a proportion of correct responses in a learning experiment (see CONDITIONING). Habits established by partial reinforcement extinguish less quickly than habits established by continuous reinforcement (see REINFORCEMENT SCHEDULE).

Pavlovian conditioning See Conditioning; Reflex

Peckhamian mimicry See Mimicry

Peck order, pecking order Older terms for what today is more often referred to as rank order. Social hierarchy of the rank order sort in communally living animals was first remarked in observations of domestic fowl, in which rank is in fact determined by who can peck whom. The terms are not quite obsolete, but are more likely to be used in a colloquial than in a technical context. The same idea was sometimes expressed by the term "peck right hierarchy."

Perception In human psychology, perception is distinguished from sensation as a higher-order capacity, sensation being the registering of stimulation and perception involving some degree of interpretation or categorization. A sensation, such as feeling giddy, need not refer beyond itself—need not, that is, be of or about something, but a perception is always of or about something, always has a content (see intentionality). Thus we perceive objects or patterns; we never merely perceive in the way that we may merely ache. There can be discrepancies between what is perceived and the stimulation giving rise to it, as exemplified by the Muller-Lyer illusion, in which lines of the same length appear to be different. In some cases such discrepancies can be attributed to differences in experience.

216

The distinction between perception and sensation is less clear when applied to animals, especially those with sensory systems different from the human. However, there is evidence that mammals and birds are subject to some of the same visual illusions that we are, such as the Muller-Lyer, and that their perceptions can be affected by what they have been led to expect or are trying to find (see SEARCH IMAGE). Study of animal perception concerns the kinds of information an animal's sense organs supply and how the animal acts on this information. Von Uexküll (1957) emphasized that an organism's phenomenal world is bounded by the kinds of stimulation to which it is receptive.

Perceptual modality   See Modality

Periodicity   Rhythm; cycling; regular repetition of some process or event. In biology periodic changes in the physical environment, such as the diurnal and tidal cycles and the rotations of the seasons, are obviously very important. Within organisms changes recur in many functions at more or less constant intervals, such as body temperature, liver function, and reproductive state. Such physiological cyclicity may depend upon internal processes, some of them within individual cells, synchronized to relevant environmental conditions by a zeitgeber. Exactly what determines endogenous rhythms has yet to be elucidated in most cases. The periodicities most in evidence in organisms are those corresponding to diurnal lunar, and annual rhythms (see ANNUAL PERIODICITY; CIRCADIAN RHYTHM).

Peripheral   Of the surface of the body. For example, peripheral stimulation is that which comes from outside and impinges on exteroceptors, as opposed to central or internal stimulation, which occurs inside the body and affects interoceptors.

Personal recognition   See Individual recognition

Phase sequence   See Cell assembly

Phase specificity   A term recently applied to cases where the effects of environmental influences on behavioral development are especially lasting when they occur during specific age stages, which are referred to as sensitive periods. Developmental processes manifesting phase specificity are socialization and all forms of imprinting. Phase specificity in this sense is recognized in human development.

Phenology The timing of recurring natural phenomena, and its study. In ethological contexts phenology refers mainly to the timing of breeding seasons, especially in relation to cyclic changes in weather conditions.

Phenotype The manifest characteristics of an organism. Phenotype is the outcome of interplay between the organism's hereditary endowment (see GENOTYPE) and the environmental and experiential conditions affecting its development (see EPIGENESIS). Differences in environmental conditions can lead to different phenotypes from the same genotype, and similar phenotypes can arise from different genotypes; for example, in a situation of complete genetic dominance, the homozygous dominant is indistinguishable from the heterozygote.

Phenotype matching A means of kin recognition by which an animal uses its own characteristics or those of kin with which it has associated to distinguish relatives from nonrelatives. It enables an animal to tell a relative that it has not encountered before from a nonrelated stranger. In the case of sweat bees (*Lasioglossum zephyrum*) it has been shown that a female guarding a nest will admit strangers as long as they are of the same family as the bees with which the female was raised, even if these were not her own siblings.

Pheromone Also sometimes ectohormone or social hormone. Pheromones are hormonelike substances produced by glands that, in contrast to true endocrine glands, secrete to the exterior of the body. They thus affect not the individuals producing them but other individuals of the same species. Pheromones fall into two functional categories, only one of which bears close analogy to hormonal function:

1. The pheromones that are comparable to true hormones are those that have a more or less directly physiological effect on recipients. Perhaps the best-known example is the "queen substance" secreted by the hypertrophied supramandibular glands of honeybee queens,

which inhibits development of other queens in the same hive. Phero-mones are important in regulating the caste composition of other social insects as well; in some cases, as in many termite species, the effects involve a complex interplay of several different pheromone substances. In some mammals the urine of a strange male can cause a pregnant female to abort (Bruce effect), and synchronization of the estrus cycling of females housed together has similarly been attributed to a pheromonal influence (Whitten effect).

2. The other class of pheromones provides chemical signals or signs in social communication, which act as releasers or as guides informing about such things as direction to food or to a sexually receptive conspecific. This category includes sex attractants, scent for marking territories (see MARKING BEHAVIOR), and alarm sub-stances, such as those discharged by various species of schooling fish and by harvest ants when they are injured or in danger (see WARNING BEHAVIOR).

Sometimes the same pheromone can serve more than one func-tion. For example, the queen substance of honeybees, in addition to its effect on female larvae, acts as a sex attractant in the nuptial flight.

Phobia   Aversion. An extreme and uncontrollable repugnance for a specific kind of object or situation. The study of phobias, which is primarily a concern of human psychology, has engaged the interest of ethologists mainly when it has to do with fear reactions in natural "situations (fear of snakes or spiders), and where, as even in some human cases, the abhorrence appears to be innate.

Phobotaxis   See Taxis

Phonology   See Sound production

Phoresis   See Hitchhiking

Photoperiod   Day length. In the higher latitudes the number of hours of daylight varies with the season, being longest at the summer solstice, shortest at the winter solstice, and equal to the number of hours of darkness at the spring and autumn equinoxes. This cycle is the most common zeitgeber for circannual rhythms (see ANNUAL PERIODICITY).

Photoreceptors Body sites specialized for sensitivity to light that respond to optical stimulation by generating afferent excitation. In most species photoreceptors are cells within special sense organs, the eyes, which may include accessory components such as lenses and pigment shields to concentrate the light on the receptors and limit the direction of receptivity. Photoreceptors may vary in spectral sensitivity, being "tuned" to different wavelengths, as are the three classes of cones, the basis of color vision, in some vertebrate eyes. Other kinds of photoreceptors are more or less equally sensitive to light frequencies over a broad range of the spectrum and give graded response according to the intensity of the light, as do the rods in the retinas of vertebrates.

In some lower animal species, photoreceptor cells are scattered over large areas of the body surface, providing "skin vision." In some protists, such as the flagellate *Euglena*, the photoreceptors are organelles—functionally differentiated entities within a cellular unit. Most photoreceptor organs register only light intensity and therefore do not allow perception of objects at a distance. Eyes providing true pattern vision have evolved in only three of the animal phyla: arthropods, cephalopods among molluscs, and vertebrates.

Phototaxis See Taxis

Phylogeny Phylogenesis. Evolutionary history or genealogy. The origin and ancestral succession constituting the evolutionary descent of a species, order, or class; a family tree of either of the two organic kingdoms—plants and animals—or of their diagnostic characteristics (see CLADISM). The phylogenetic study of behavior follows the same principles as that of other characteristics of organisms. As in tracing the evolutionary origins and transformations of structures, behavioral phylogeny involves identifying and tracing homologies, a comparative morphological approach. In ethology this approach has been most successful in work on the evolutionary provenance of displays, which are viewed as derived activities (see RITUALIZATION).

The main influences governing the course of phylogeny are mutation and selection. Distinct from phylogeny or phylogenesis is the development of the single individual, which is referred to as ontogeny or ontogenesis.

Physiological system   A complex of organs and tissues responsible for one of the major biological functions of the body, for example, the reproductive system, blood-vascular system, central nervous system, or respiratory system.

Physiology   Study of how organisms work; the functions and processes of the body that constitute life and living. They include metabolism, motor mechanisms, hormonal regulation, and the reception, transmission and processing of stimuli. The divisions of physiology are designated according to their focus: sensory physiology, neurophysiology, endocrinology, and cellular physiology, for example. The physiological basis of behavior is thus called behavioral physiology.

Piloerection   Erection or bristling of hair. The functions of piloerection are analogous to those of feather ruffling in birds, namely temperature regulation and visual signaling. However, the visual releaser function appears to be less widely distributed and less strongly marked in mammals than in birds, which goes with their generally more modest coloration. As with feather ruffling, piloerection can occur as a side effect of physiological arousal.

Pituitary   Hypophysis. An endocrine gland hanging from the floor of the brain stem in the forebrain region in vertebrates. It is sometimes referred to as the master endocrine gland of the body, for its secretions regulate the activity of several other endocrine glands situated elsewhere. The pituitary produces two knds of hormones: those that affect the body directly, such as prolactin, which induces mammary development, among other things, and glandotropic (glandotrophic) hormones, which influence the activity of other endocrine glands, including the thyroid (via thyrotropic hormone), adrenal cortex (via adrenocorticotropic hormones, or ACTH), and gonads (via gonadotropic hormones), and so produce indirect physiological effects (and perhaps some direct effects). Pituitary hormones affect behavior both directly and indirectly. Prolactin, for example, can sustain incubation behavior in birds by acting on the operative sites in the brain, while the gonadotropic hormones influence reproductive behavior by the roundabout route of their influence on the hormonal secretion of the gonads.

Secretion of hormones from the pituitary is under the control of the overlying hypothalamus. For the anterior lobe the connection is via a system of capillary blood vessels called the hypothalamohypophysial portal system. Nerve cells in the hypothalamus secrete chemical agents called releasing factors (or hormones) into this portal system, which cause the endocrine cells in the anterior pituitary to discharge their hormones (in the case of the prolactin-secreting cells the hypothalamic factor inhibits hormone release and is therefore called the prolactin-release-inhibiting factor). The hypothalamus communicates with the posterior part of the pituitary (neurohypophysis) via looping axons from hypothalamic neurons. The posterior pituitary hormones, such as oxytocin (which in mammals induces the uterine contractions of parturition and the ejection of milk during suckling), are manufactured in the hypothalamic cell bodies of these neurons, carried down the axons into the gland, and there released into the systemic circulation when the cells receive the signal for secretion. In the animals possessing it, the hypothalamohypophysial system constitutes an "interface" between the two main regulatory systems: the central nervous system and the endocrine system.

Place imprinting Home imprinting; site imprinting; locality imprinting. Fixation on a particular geographical region as a consequence of early experience. It is an important prerequisite for the many animal species that show fidelity to place. Examples of place imprinting are familiar from migratory fishes and birds. Thus Pacific salmon imprint at an early age on the chemical characteristics of their home stream and so are able to find it when later returning from the sea to spawn ("olfactory imprinting"). Shortly after leaving the nest, American indigo buntings learn the star patterns in the night sky, which subsequently guide them on their return from their wintering grounds. Place imprinting is like other imprinting phenomena in the early and often very brief sensitive period during which the habitat characteristics are learned and in the persistence of the preference.

Placenta   A gestational structure that is diagnostic of the eutherian (placental) mammals, in contrast to the monotremes and marsupials. The placenta makes an anchorage in the wall of the uterus and attaches to the fetus via the umbilical cord. It takes various forms, but in all it provides the means of physiological communication between the fetus and the mother. In the placenta the fetal and maternal blood systems are brought into intimate proximity through an intermeshing of embryonic and uterine wall tissue, so that the fetus's nutritional, respiratory, and excretory needs are met by the mother via the bloodstreams. In addition the placenta serves as a temporary endocrine gland (see PROGESTINS). It is shed as part of the afterbirth at parturition.

Play behavior   A category of behavior that defies precise definition. Perhaps the best suggestion is that play is behavior performed without the "serious point" that such behavior has in its normal context. Play differs from serious behavior in a number of respects. It consists either of newly invented movement patterns or of whole or part patterns having serious purpose in other contexts, such as attack, escape, or prey-catching actions, but which appear quite senseless in the play context because the animals hold back from following through. Moreover, such actions are often combined with others they would not normally occur with; they may lack the usual constraints of reciprocal inhibition; and their sequential patterning may differ from the more or less fixed order of action chains. Play behavior appears to be free of the habituation that may affect the behavior in its serious setting, since it can sometimes be repeated continually for long periods without showing any signs of waning. In social contexts roles may change frequently and almost arbitrarily. The behavior sometimes appears "overdriven," that is, it is performed in more exaggerated form, with greater expenditure of energy, greater rapidity, and more repetition than normally. Play usually includes numerous spontaneous components.

Play behavior occurs only in the more highly evolved vertebrates—the birds and mammals. As a rule it is restricted to young animals, although in certain cases, especially in carnivores, rodents, primates, and whales (most notably dolphins), it persists to some degree in adults as well. In accordance with object relations, we can distinguish between object play, in which a lone animal sports with an inanimate object or toys with a creature of another species (a cat playing with a mouse), and social play involving two or more conspecifics or individuals of different species that treat each other like conspecifics, as some domestic cats and dogs do. A kind of play behavior in the acoustic domain is the subsong of many bird species, which can be compared to some extent to the inventive acrobatics of brachiating primates.

The biological significance of play is a topic of frequent discussion in the ethological literature. The most likely possibilities are that it contributes to motor development, for example through exercise of muscular function and acquisition of manipulative skills; to cognitive development, through exercise and improvement of perceptual capacities; and to social development, through practice in social roles and individual recognition of social companions, and development of social communication. Also it may differentiate young from older animals and thus function as a juvenile behavioral characteristic. Altogether more than thirty different functions for play have been suggested in the literature. Most are consistent with the general statement that the main biological significance of play is in enabling an animal to acquire confidence and skill in the use of its body, to discover what it can do and how to cope with the world in both its physical and social aspects by trying things out and imitating.

On the question of the motivation underlying play behavior there is as yet no clear answer. The differences between play and serious forms of the behavior make it unlikely that the motivations are the same. There are various indications of an "appetite for play" (see APPETITIVE BEHAVIOR), meaning that play behavior may have its own motivational basis—a "play drive." But conclusive proof of this is lacking. Some animals have developed special signals for the play context (see PLAY SIGNAL).

Play face   See Play signal

Play signal   A facial expression or gesture that signifies readiness for social play (see PLAY BEHAVIOR). Whether such signals are effective as releasers is still a matter of doubt. Play signals serve also for the avoidance of "misunderstandings," especially during aggressive play, when a combatant may signal that certain of its aggressive actions are not seriously intended. For those species in which play has evolved its own signals, play behavior has apparently attained considerable biological significance. The best-known signal is the "play face" of many primates and carnivores, which is probably a ritualized intention movement of biting.

Play song See Subsong

Pleiotropism See Gene

Polyandry See Polygamy

Polyethism Behavioral polymorphism. The regular occurrence of alternative forms of behavior within a species or population. The behavioral options may be either interchangeable, as in the role taking of many birds and mammals (see ROLE; STRATEGY), or allotted for life, in which case they are usually attended by appropriate morphological differences, as in many social insects (see CASTE; SOCIETY). Honeybee workers carry out a more or less fixed succession of activities, inside and outside the hive, during the course of their lives. Polyethism is the basis of true division of labor.

In addition to alternative forms of manifest behavior, polyethism refers to the establishment of preferences, especially concerning ecological conditions, as in attachment to a specific habitat type. The process often involves the effects of early experience (see HABITAT IMPRINTING).

Polygamy A mating system in which individuals of one sex couple with more than one of the other sex. The arrangement in which a male takes several females is called polygyny, and that in which a female consorts with several males is called polyandry. Polygyny is quite common, but polyandry is very rare; it is known only in several species of birds, such as jacanas and the Tasmanian waterhen. Unlike monogamy, polygamy occurs mainly in species in which one sex (the one that has multiple mates) takes no part in parental care, except in special circumstances; therefore that sex has the time and energy to defend several partners and sufficient territory to maintain them and their offspring. Accordingly, in contrast to the monogamy that prevails among birds, polygamy is the most widespread mating system of mammals, in which the females have to bear and nurse the young and so "automatically" have the major part in parental care. However, in some fish, such as the three-spined stickleback, the males perform multiple matings as well as all of the parental care.

With regard to duration of relationship, polygamy presents two patterns: in many species, such as most weaverbirds, the polygamous sex mates with a succession of partners without forming any lasting bonds (see PROMISCUITY); in other cases, as with many mammals and fish, a male retains possession of a group of females for an extended period, varying between a season and a lifetime. Such an arrangement constitutes a harem.

Many songbirds—wrens and red-winged blackbirds, for example—have a system that has been described as "facultative polygamy": a mated male will take on a second female or even more, apparently as circumstances allow; or a female choosing a mate may have to choose between an already mated male with a high-quality territory and an unmated male with a low-quality territory.

Polygene See Gene

Polygyny See Polygamy

Polymorphism The regular occurrence of alternative forms within a species. It can be either genetic or phenotypic. In genetic polymorphism two or more gene forms (alleles) occur at the same chromosomal locus in frequencies greater than mutation rate or immigration would produce. Such polymorphism can be temporary, lasting only so long as it takes for selection to exclude all but the fittest alternative; or it can persist as a balanced polymorphism, as when heterozygosity confers greater fitness than homozygosity. An example of the latter is the persistence of sickle cell anemia in human populations living in places where they are subject to malarial infection: in the heterozygous condition the sickle cell allele confers resistance to malaria, which counterbalances the deleterious effects it has in the homozygous condition. An example with behavioral consequences is the inheritance of the T-allele in wild house mice (*Mus musculus*): the homozygous combinations of T-alleles are lethal or semilethal, but males having a heterozygous combination

226

have a reproductive advantage in competition with homozygous males.

Nongenetic or phenotypic polymorphism differs from other forms of intraspecific variability in that discontinuities separate the different forms, that is, the different phenotypes do not grade into one another via intermediate forms. Such polymorphism is widespread in the animal kingdom. It is especially prominent in the social insects. Alternative forms of behavior constitute cases of behavioral polymorphism or polyethism.

Population   A collection of animals of the same species occupying a more or less sharply bounded area within which interbreeding maintains continuous gene flow and hence mixing of the members' genetic material. A similar but slightly more technical concept is that of the deme: a localized, bounded population within which breeding is completely random (panmictic). Such a situation is probably never fully realized in nature, but the idealized concept is assumed in some of the theoretical models used in population genetics. The interbreeding population is the basic unit of evolution. As a rule a population is isolated geographically, and hence reproductively, from other populations. However, the geographical barriers between such allopatric populations sometimes break down, and then at least parts of the population ranges may overlap. Where that is the case the portions in contact are said to be sympatric. In sympatric populations, where hybrid matings are less successful than matings within a population, selection can result in the evolution of other forms of reproductive isolation, such as discriminative pair formation. There may even be preadaptation to this end, as in the cases of bird populations that have local song dialects, predisposing the birds to pair with their own kind. The evolution of behavioral barriers to interbreeding can thus play a part in speciation.

Population biology   That part of biology concerned with describing and analyzing the temporal and spatial distributions of the members of a population and their relationships to all the relevant living and nonliving factors of their environment. Population biology, including population genetics, makes important contributions to the research programs of sociobiology.

Population density The average number of individuals of a population per unit area. Density is affected by a number of environmental factors, such as food supply, predator pressure, and competition from other species for limited resources. Its limits are characteristic for each species. As dictated by food supply, the population is, as a rule, less dense for carnivores than for herbivores, and less dense for larger than for smaller animals. More or less regular fluctuations of quite considerable magnitude are characteristic of some species. Many of these have an annual periodicity, but some involve longer periods. The Northern lemming is a famous example of an animal showing population density cycles with several years between peaks.

Excessively high population densities are known to cause stress phenomena. Some species have evolved modes of population control, including regulation by social means, that serve as a buffer against the direct effects of resource availability (see GROUP SELECTION).

Population ecology See Ecology

Population genetics Study of genetic flux and the factors affecting it in populations. It deals with such matters as mutation rate, selection pressure, genetic drift, and population density, and attempts to relate them to one another mathematically in theoretical models that account for known facts and predict what the facts should be in untested cases.

Positive feedback See Feedback

Postcopulatory behavior Behavior occurring immediately after copulation. Its function is still a matter of doubt and probably varies from species to species. In the majority of cases the behavior appears to allay renewed aggressive or fleeing tendencies, thus allowing the partners to stay together (see PAIR BONDING). If this is an accurate interpretation, postcopulatory behavior is a type of appeasement behavior. In the Uganda kob, the males form an assemblage with adjoining courtship arenas (see LEK). They have a pronounced postcopulatory display, apparently to induce the females, which are capable of multiple copulations, to remain and to discourage them from moving to the arena of a neighbor. The old idea that postcopulatory behavior discharges surplus energy left over from coition has yet to be substantiated.

Behavior immediately after copulation is also sometimes called postcopulatory play or copulatory afterplay. However, since it does not appear to be play behavior in the strict sense, the term "post-copulatory behavior" is preferred.

Postural facilitation   See Transitional action

Potential   This term occurs in physiological and behavioral contexts. In physiology it refers to a difference or change of electrical state. For example, the electrical difference between the inside and outside of a nerve fiber is referred to as the membrane potential. The inside of a cell at rest is about 70 millivolts negative with respect to the outside, and this is called the resting potential. The transient propagated change in membrane potential that constitutes a nerve impulse is called an action potential.

In behavioral contexts "potential" or "potentiality" means much the same as "tendency"—the probability that action of a certain sort will occur.

Pragmatics   See Communication theory

Preadaptation   A characteristic that proves to be of adaptive significance in circumstances different from those in which it originally arose and in which it was of indifferent importance. A behavioral example is that of song dialects, which initially develop through geographical isolation but which can provide a means of achieving reproductive isolation through discrimination during pair formation when the populations later become sympatric and selection discriminates against hybrid mating.

Preceptive behavior   See Soliciting behavior

Precocial animal   A species in which the young are at a relatively advanced stage of development when born and are thus able to follow their parents after only a short time. The term is used mainly of species in taxonomic classes that also include the contrasting category of altricial animals, such as birds, mammals and fish. Altricial

229

young emerge with only the rudiments of sensory, motor and thermoregulatory capacities and so require considerable parental care for the provision of food, warmth, and protection during the first days or weeks after hatching or birth. Sometimes ant species are described as precocial, as in the South American army ants, in which workers participate in the colony's treks and raids very soon after emerging from the pupal stage.

Distribution of the two developmental types generally follows taxonomic lines; that is, in most cases orders within the class Aves and the class Mammalia are consistently either one or the other. Thus in birds all ostriches, gallinaceous birds, ducks, and geese are precocial, as are all ungulates among mammals. However, the rodents and the lagomorphs contain both precocial (guinea pigs, hares) and altricial species (rats, mice, rabbits). In birds the precocial condition is apparently phylogenetically the older; conversely, in mammals it is the more recently evolved, "more progressive" development. Both vertebrate classes show an evolutionary trend from larger to smaller broods or litters and they differ correspondingly in that the precocial species in birds and the altricial species in mammals have the larger broods or litters.

Besides the phylogenetic developmental level, ecological adaptation appears also to have played a part in the evolution of the two types of offspring. In particular, predation pressure can induce precociousness in species living on open ground, as is quite evident in groups in which both types occur in closely related species. Thus the ground-nesting cranes and surface-living hares are precocial, and the tree-nesting storks and burrow-breeding rabbits are altricial. A kind of intermediate condition between precocial and altricial is presented by the clinging young of primates and some other mammals, and also by bird species whose chicks hatch with open eyes, down feathering, and enough motor control to stand and walk but are nevertheless dependent on their parents for food and warmth. Gulls are typical examples. Such birds are sometimes described as semiprecocial (compare NIDIFUGOUS NESTLING).

Preening   Care of the plumage. Birds use their bills to clean, comb, and card their feathers, smoothing them out after ruffling, rehooking detached barbules, and in some species dressing the feathers with secretions from an oil gland between the wings. Scratching and head rubbing movements may also be included in preening. Bouts of preening often show regular sequential patterning that is characteristic of either the species or the individual. Such bouts are frequently associated with other forms of comfort behavior, such as bathing and dusting. Like grooming in mammals, preening movements often occur as displacement activities.

Preference   Ecological writing often refers to specific features of the physical and biological environment that individuals or species are drawn to when there are conceivable options, such as particular temperature ranges, humidity levels, light intensities, foods, and habitats. For ethology, social choices are of prime interest, for example an animal's fixation on a particular category of individual as a consequence of sexual imprinting and its selective response to particular individuals, which opens the way to study of the capacity for individual recognition. Preferences are routinely investigated by means of choice tests. Under natural conditions individual preferences may lead to Darwinian sexual selection.

Prenatal learning   Learning that occurs prior to birth or hatching. In some species of birds active acoustic communication between a chick and its mother (and between chicks) may develop even before hatching from the egg. The chick can thus acquire the ability to differentiate between known and unknown sounds or between sounds associated with affective change and sounds lacking such contingency. When certain recorded sounds are played in an incubator containing eggs of domestic chicks, the chicks, on hatching, prefer those sounds to previously unheard sounds. In the guillemot, an auk that breeds on cliff ledges, the parents reply to the calls of their unhatched young, and the young thus learn to recognize their parents' calls and subsequently to distinguish them from those of neighbors. This special learning capability is an adaptation to the very dense breeding colonies of guillemots, in which pairs incubate their eggs "shoulder-to-shoulder," without nests. It ensures that parents and young recognize one another at hatching and hence that the young can be raised without confusion.

Prenatal learning in mammals, whose embryos are considerably more insulated from outside influence, has yet to be definitely established.

Preparedness See Learning disposition

Presentation A term used in the ethological literature for an animal's display or "showing off" of specific body parts or appendages to a conspecific. The best-known example is the presenting of hindquarters by primates, which has a double social function: it serves as female sexual soliciting behavior and as an appeasement gesture of both males and females toward a higher-ranking group member. "Weapons" are often presented as a part of threat behavior, as in the horn presentation of antelopes.

Primary sex characteristics See Sex characteristics

Primatology Study of primates, the order of mammals comprising the prosimians (bush babies, lemurs, tree shrews), monkeys (New World platyrrhines, such as marmosets, howlers, and spider monkeys, and Old World catarrhines, including macaques, baboons, colobus monkeys, and vervets), apes, and human beings.

Because of the close phylogenetic relationships, investigation of the behavior of nonhuman primates can contribute to understanding of the foundations of human behavior. A much-cited instance is research on the deprivation syndrome, carried out first on rhesus monkeys and later extended to other primates, which has led to new insights into the needs of human infants for social attachment. Primate studies are included in some areas of anthropology for the same reason.

Priming stimulus Motivating stimulus. A stimulus that has no immediate overt effect on an animal's behavior but that nevertheless increases the response tendency, as manifested by reaction to subsequent stimulation. For example, female mice show only weak parental response when presented with a dead pup, but if the presentation is preceded by exposure to a living pup, the mouse reacts to the dead one much more strongly. The first stimulation thus primes the response of the animal to the second. Such an influence can persist for several days. Covert increase of motivation through threshold change is also referred to as sensitization.

Problem solving   Studies of animal intelligence frequently include testing of animals on problems that can be solved by inference or insight as opposed to trial-and-error learning. For example, when an animal is familiar with a particular maze, the path it has learned to the goal box is blocked off to test whether its knowledge of the maze can enable it immediately to take a detour (see COGNITIVE MAP). Choice tests of various sorts have been used to find out the extent to which animals register relationships such as "larger than," form concepts such as "threeness" (see CONCEPT FORMATION; COUNTING ABILITY), recognize oddity, or match to sample. Primates have proved to be the most adept at such problems. In natural situations animals may encounter problems that call forth intelligent response, such as a practical invention that leads to tool use.

Proceptive behavior   Sexual invitation by female mammals to solicit copulation from males (see SOLICITING BEHAVIOR). The genital presentation of female primates is a prime example.

Progenesis   See Neoteny

Progesterone   See Progestins

Progestins   A group of female sex hormones produced in the ovary by the corpus luteum and by the placenta in mammals and by analogous (if not homologous) tissues in other vertebrates. Progestins are involved in the regulation of reproductive processes, but as a rule later in the cycle than estrogen. They influence some aspects of reproductive physiology and behavior subsequent to copulation, such as preparation of the uterine lining for implantation of the fertilized ovum in mammals and induction of incubation behavior in birds. The best-known progestin is the luteal hormone progesterone.

Program   The terms "program" and "behavioral program," in their technical senses, are recent additions to the ethological lexicon, drawn from the language of computer science to help deal with the recurrent and controversial issue of hereditary or innate components of behavior. The control of behavior is compared to the governing of data processing in a computer. The program circuitry either is built into the structure of the machine as a permanent feature of the way it works (so-called hard-wiring or hardware), or is set up by computational instructions entered into memory storage for a particular case and therefore variable from one case to another (soft-wiring or software).

Analogously, control of behavior in animals depends upon the information structuring the central nervous system or temporarily stored there, which governs the manner and flexibility with which behavior adjusts to environment. Such adjustment presupposes that the organism has relevant information about the environment, which can be supplied in two ways: through heredity and through experience. Information may be stored in the genome—represented in a "species memory," as it were—and so transmitted from one generation to the next; or it may be acquired *ab initio* by individuals, and so deposited in the memories of single animals separately. In conformity with computer language, the aggregate of information from both sources governing a behavior pattern is referred to as a (behavioral) program. Information of the first sort, which is built into the organism's "computer," corresponds to the hardware of the electronic computer, while information acquired through individual experience, which can mediate adjustment to new situations, corresponds to software. When the behavior of a species is governed predominantly by hereditary information, as is probably the case with primitive groups such as the lower invertebrates, it is described as being governed by a closed (behavioral) program; in contrast, when individually acquired information plays a modifying or determining role, an open (behavioral) program is ascribed. The contributions of the two types of program may differ not only between species but also between different behavior systems within a species.

There are two important differences between the animal and the computer situation. First, the genetically guided neural structures and behavior patterns always develop in close interplay with the environment (see EPIGENESIS). Consequently organisms do not have hardware in the literal sense of a computer. Second, in animals, programs that are genetically transmitted as "open" often become so firmly fixed through individual experience in early life that they can no longer be changed at later stages, so they work essentially like closed programs. This applies most obviously to the behavioral characteristics determined by imprinting. Here, however, the hardware originates from input consequent on initial operation, as well as from

234

the program in the structure of the "computer." Because of the varying degrees of fixity in the effects of imprinting processes (see IRREVERSIBILITY), the distinction between the two categories to which the computer terms are attached may become fluid. The reason for this difference lies in the fact that a computer begins to work only after its manufacture is complete and its hardware installed. A young organism, in contrast, must be able to process information *before* reaching final form; consequently it builds new hardware after starting work. This "vagueness" of the boundary between hardware and software in living organisms is one reason why the use of the traditional "innate-learned" antonymy, which suggests a hard and fast dichotomy, has so often been rejected in the history of ethology.

When an organism has at its disposal several programs for a single behavior system, which can be selected according to the situation, ethologists may refer to the animal's "strategies". The main difference between the uses of the terms "program" and "strategy" in behavioral study is that in the investigation of programs the concrete *realization* of strategies is emphasized (*how* the organism develops a specific strategy), while in the analysis of strategies this question is irrelevant. See also SOCIOBIOLOGY.

Prolactin   Lactogenic hormone. A hormone produced by the anterior pituitary that promotes mammary development and milk production in mammals and crop-milk production in pigeons and doves. It may also be involved in the development of brood patches and the sustaining of incubation in some birds.

Promiscuity   A kind of mating system in which the sexual partners do not form persistent pair bonds, meeting only for the purpose of copulation. One male may mate with several females, or a female with several males. The boundaries between promiscuity, polygamy, and harem arrangements are blurred. Thus in species with lek display arenas, the group stays together only for the time required for copulation and its preliminaries; in other species, as in many seals, the sexes stay together for some days or weeks, during which partnerships may change frequently. As with polygamy, species living promiscuously usually display pronounced sexual dimorphism, and only one sex, usually the female, is responsible for parental care.

Proprioceptor   An interoceptor that registers the position and movement of the parts of the body relative to one another and thus gives an animal information about its posture and its active and passive movements (see KINESTHESIA). Proprioceptors are found, for example, in the muscles and tendons of vertebrates (muscle spindles, Golgi tendon organs), and arthropods (stretch receptors, chordotonal organs).

Protean behavior   Escape behavior in which a pursued prey animal switches unpredictably between different fleeing or avoidance patterns. For example, a moth under attack by a bat will go through a series of erratic flight maneuvers to baffle the bat's attempts to track and grasp it. Another form of protean behavior consists of sudden change of appearance, as when insects approached by birds produce simulations of frightening predators. The behavior is named for Proteus, a minor deity in Greek mythology who had the power to assume any shape he wished.

Proximate cause   Proximate factor. A general term for factors, such as external stimulation and motivational state, governing the immediate causation of a physiological process or pattern of behavior. For example, in many bird species the tactile stimulation of sitting on eggs leads to cessation of egg laying. Thus the timing of the onset of incubation is part of the proximate cause determining clutch size in these birds. Contrasting with an account of *how* clutch size is determined is *why* the clutch is the size it is in terms of the number of young that can be effectively provided for, which relates to food availability as an ultimate cause.

   A similar contrast can be drawn between the proximate and ultimate functions of a behavior pattern: the immediate effects or consequences of the action and the ways in which such effects or consequences contribute to survival or reproductive success. For example, an immediate function of birdsong is threat and hence territorial defense; an ultimate function may be the spacing of individuals to ensure adequate food supply for successful breeding.

Pseudoconditioning   Increase in the probability that a previously neutral stimulus will elicit a particular response as a consequence of repeated elicitation of the response by the stimulus it is already associated with, but without the contingent relationship between the two stimuli that obtains in true conditioning. The animal may generalize from the initially effective stimulus to the neutral stimulus, or the repeated elicitation of the response may change the animal's state in such a way as to increase the likelihood that any stimulus will evoke the response, as when repeated electric shock causes an animal to avoid any stimulus, whether associated with the shock or not.

Pseudoconditioning is different from sensitization, which is the potentiating of one stimulus-response connection as a consequence of repeated activation of another. If the two responses are similar in form, this effect, like that of pseudoconditioning, may look like true conditioning, but again the required contingency relationships are lacking, and the effect can be attributed to a change in general reactivity. Scientists working on animal learning often have to include special control groups in the design of their experiments to rule out the possibilities that pseudoconditioning or sensitization are responsible for what appears to be learning.

Pseudocopulation   A form of copulation that lacks the function of fertilization. It has been described in fishes, birds, and mammals. In mammals it may consist of mounting with or without intromission. Pseudocopulation can take place between a male and a female, two males, or two females. In the heterosexual situation it probably serves as support for the pair bond and represents an example of sexual behavior emancipated from an immediate reproductive connection (see BONDING BEHAVIOR; EMANCIPATION). Pseudocopulation between same-sex animals is difficult to interpret. Its biological significance probably varies with different species and relationships. In primates, where it is especially common, it can be viewed as a demonstration of higher status within a rank order. However, this explanation will not do for all cases, for in many species the individual mounted may be the dominant one. In all-male groups pseudocopulation possibly serves to maintain reproductive readiness (see HOMOSEXUALITY). As a rule pseudocopulation is most in evidence during occasions of general excitement and in conflict situations. It may also occur during play behavior.

Pseudopregnancy   A condition that can occur in female mammals when mating fails to cause conception. The stimulation of copulation gives rise to the same hormonal changes that result from pregnancy and hence the other physiological changes that go with that condition, such as cessation of ovarian cycling (see ESTRUS).

Psychogenetics An infrequently used term for the genetic study of human behavior (see BEHAVIORAL GENETICS). It occurs mostly in the literature of psychopathology to refer to the possibly hereditary bases of behavioral characteristics.

Psychogenic social regulation The influence of psychological factors, such as individual attachment, emotional investment, and relationships of dominance and subordinance, in the control of behavior in groups. The concept was introduced by Schneirla (1949) to distinguish the kind of social regulation typical of higher vertebrates from that governing the societies of social insects, which he described as subject to biogenic social regulation.

Psychological castration See Gonadectomy; Social inhibition

Psychology The study of mental life and behavior. It consists of a number of divisions whose conceptual and methodological differences are so great that some authorities have written of the *sciences* of psychology. Physiological psychology deals with the machinery of the body insofar as it is relevant to psychological functioning. Experimental psychology uses the resources of the laboratory to pursue questions about such matters as psychophysics (the relationships between the magnitude of stimuli and their experienced intensities), the principles of learning, and the nature of motivation. Cognitive psychology deals with perception, memory, and thinking. Developmental psychology deals with how behavioral and cognitive capacities develop in children and young animals and how they are affected by aging. Social psychology examines the behavior and relationships of people in groups. Abnormal psychology is concerned with the description, explanation, and treatment of neuroses, psychoses, and other mental and behavioral disorders. The psychology of personality has to do with variations in personality types and how they arise, especially in the light of psychogenic theories such as those of psychoanalysis. Among the numerous forms of applied psychology are industrial, educational, vocational, and clinical psychology. To varying degrees these kinds overlap with one another and with neighboring sciences such as sociology and anthropology. By and large psychology is much more concerned with human beings than with animals, and when it uses animals for experimental purposes it is often for the sake of arriving at conclusions bearing on the human case. However, ethology has much in common with comparative psychology and also draws from physiological and experimental psychology.

Psychopharmacological drugs Chemical substances that affect the central nervous system and lead to changes in behavior. In human beings psychopharmacological drugs are commonly used in studies of memory, regulation of reproductive behavior, and many other behavioral phenomena. For example, analeptic drugs such as strychnine have been found to affect short-term memory, and certain antibiotics have been found to affect long-term memory.

Punctuated equilibrium A term introduced by Eldredge and Gould (1972) to characterize a theory about rates of evolution. It opposes a more conventional view, stemming from Darwin, according to which microevolution, such as occurs in populations, and macroevolution, such as causes the origin of species, are on a continuum, with the same gradual sifting of heritable variations by selection applying throughout. This view is described as phyletic gradualism. According to the theory of punctuated equilibria, there is a radical difference between microevolution and macroevolution. It holds that species are established by sudden bursts of change, in terms of the geological time scale, and then settle into a relatively static history until they become extinct. Instead of individual or genic selection governing evolution within a species, the origin and extinction of species that constitute macroevolution are thought to involve some form of species selection. These ideas are highly controversial as regards both the proposed pattern and the proposed process of evolutionary change.

Puzzle box A general term for any cagelike container used for the investigation of learning, memory, and other cognitive capacities of

239

animals. Examples are the maze, the Skinner box, and the Wisconsin General Test Apparatus. Credit for initiating the use of puzzle boxes in the study of animal intelligence is generally given to the American psychlogist E. L. Thorndike (1911).

Quantitative ethology   A term sometimes used for the quantitative recording and subsequent analysis of behavior. It provides an important basis for the compilation of an ethogram. The quantitative treatment of characteristics that allow recognition of bonding within a social group is referred to as sociometry.

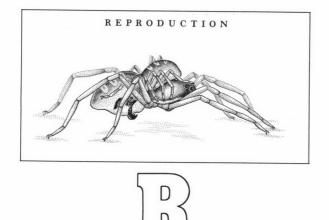

REPRODUCTION

R

Rank mimicry   Imitation by lower-ranking individuals of the behavioral characteristics of higher-ranking individuals (see RANK ORDER). In primates it may play a role in dependent rank (see BASIC RANK). An acoustic example of rank mimicry has been noted in some birds, in which young males selectively learn the songs of higher-ranking males. So far this tendency has been established only in polygynous and promiscuous species, in which it probably aids the mimicking males in their efforts to attract females.

Rank order   Social hierarchy, dominance hierarchy, peck order. Ordered distribution of "rights and obligations"—precedence and subordinance—within a group of animals living together. The stronger or more experienced individuals stand higher in the order and consequently enjoy certain privileges, such as first access to resources. Social gradation of this sort is predominantly found in vertebrates (fishes, reptiles, birds, and mammals all provide examples), but sporadic occurrences are known from invertebrates (insects) as well. Its form varies considerably from species to species. In many cases, for example in the domestic chicken, rank order is

241

linear. In other cases more complex orderings occur, such as triangular relationships, in which one animal is dominant over a second and the second over a third, but the third has ascendancy over the first. Further, as is quite strikingly the case in many primates, the rank position of an animal may vary in different behavioral contexts. The rank order may include both sexes or be separate for each sex. In the latter case a member of a pair sometimes has the same rank as its partner, the status usually being determined by the male (see BASIC RANK). In some primates the male ranking may be complicated by formation of a central hierarchy of two or more animals that support one another and collectively dominate other males, which may be capable of defeating any one of them in single combat. In the earlier ethological literature, at least, the position of an animal in a rank order is usually indicated by a letter of the Greek alphabet. The highest-ranking individual is the alpha animal; the lowest-ranking, the omega animal.

The highest-ranking individual is probably first in line at feeding time and at the water hole, can choose the best place to sleep or rest, and can more or less monopolize the mating opportunities. However, the position can also carry certain obligations, such as assumption of leadership of the group and readiness to act as its guardian and defender. Because of these facts, and because in many species the rank order relationships are very complex, description in terms of roles is increasingly preferred to the original ranking terminology.

Rank-order hierarchies serve to stabilize relations within a group. The only contention, as a rule, is in establishing the hierarchy and subsequently changing it when young animals grow old enough to challenge their elders; otherwise the priorities of rank are respected without contest. The development and maintenance of dominance relationships more or less necessitate individual recognition between group members.

Although rank-order relationships are typically a feature of group-living animals, they are not restricted to them. Indeed a kind of ranking can obtain between the partners of a pair. This is most conspicuous in monogamous species, especially among birds. In most such cases the male is dominant for most of the year, but at the beginning of the breeding period, when the female's parental investment is greater than the male's, the dominance relationship may be temporarily reversed. A kind of rank order is sometimes found between members of different species living in a single region and exploiting the same natural resources. It can appear as the order of species succession, sometimes established through aggressive interaction, in access to a watering place or a choice source of food, for example, among carrion eaters at a carcass. Hediger (1941) has called such regulation "biological rank ordering."

Reaction chain   See Action chain

Reafference   Sensory stimulation (see AFFERENCE) consequent on active movement of the body, hence a form of feedback. It is usually dual in nature: exteroceptors, such as the eyes, convey change of position in relation to the environment (image displacement on the retina), and proprioceptors inform about change of position of parts of the body relative to one another (see kinesthetics). Through comparison of these two inputs the animal can determine whether change with respect to the environment occurred solely because of the body's active movement or because the body was additionally or totally subject to passive movement (drifting by wind or water current). In the case of the latter, reafference enables the animal to make compensatory movements to maintain its position. According to the "reafference principle," for every command to the effector organs—the muscles pertaining to locomotion, say—an efference copy (*Sollwert*) is deposited in the central nervous system (see FEED-FORWARD). This copy is checked against the stimulation resulting from execution of the command—the reafference. If the feedback stimulation comes about only through active self-movement, it will match the "expectation" of the efference copy, thus clearing the copy from the system and allowing the animal to go on to its next action. If, on the other hand, passive movement is also involved, there will be discrepancy between the feedback and the efference copy. In this case the discrepant stimulation can be distinguished from the reafference as exafference and will elicit the appropriate compensatory action.

Reasoning   See Problem solving

243

Recapitulation In biology, the "recurrence of adult ancestral characteristics during . . . ontogeny" (Rensch 1954). Its apparent ubiquity led the nineteenth-century German biologist Haekel to formulate his "biogenetic law": ontogeny recapitulates phylogeny. However, it has been argued that much, if not all, of the evidence for recapitulation in this sense can be more plausibly interpreted as recurrence of *juvenile* ancestral characteristics during ontogeny. Most cases of supposed recapitulation are anatomical, but there are some behavioral examples: many species during juvenile development manifest transient behavioral patterns that occur in the adults of other, usually closely related species, generally the more "primitive" species within the group. It is therefore supposed that the ancestors of the "advanced" species likewise possessed the behavior concerned, that is, that the behavior has subsequently been lost except for its temporary appearance in the young (see BEHAVIORAL PHYLOGENY).

Examples of behavioral recapitulation are provided by some ground-nesting songbirds (larks and pipits), which run with alternating leg movements as adults, but hop with both legs simultaneously during the first few days after leaving the nest, conforming briefly to the behavior of closely related tree-nesting species. Also the larvae of stick insects show patterns of coordination in their leg movements that differ markedly from those of the adults, yet resemble those of cockroaches; stick insects derive from cockroachlike ancestors.

It is not always easy to determine whether recapitulation phenomena represent pure relicts, or whether the behavior in question also serves a function associated with juvenile body characteristics (size) or ecological requirements (foraging patterns in a living space different from that of the adult).

Receptive field For any unit in a sensory system, the area from which it receives input that changes its activity. The effects can be either an increase in the activity of the unit (positive input) or a decrease (negative input). For example, most of the axons (neuron fibers) in the optic nerves of a cat have receptive fields at specific spots on the retina; these consist of an excitatory center ("on" center) with an encircling inhibitory ring ("off" surround). This arrangement is a consequence of the fact that a retinal receptor cell excites the optic nerve neurons immediately underlying it but inhibits those with which it connects (via intermediate cells) in the surrounding area. This process of lateral inhibition has the effect of enhancing contrast in perception of the edges of retinal images. Units have been discovered in the visual systems of frogs that have receptive fields tuned to the visual patterns presented by flying insects. These "bug detectors" provide the kind of stimulus filtering that releasing mechanisms were postulated to provide.

Receptor Sensory site. In multicellular animals receptors are cells that register stimulation within the body (interoceptors) and stimulation impinging on the outside or periphery of the body (exteroceptors), transforming or encoding it as neural transmission to the central nervous system. Different receptors are tuned to different kinds and ranges of stimuli (their "adequate stimuli"); that is, they possess stimulus specificity and can be designated as photoreceptors, chemoreceptors, and so forth. Sense cells may occur singly (tactile nerve endings in the skin) or be combined in sense organs.

Another use of the term "receptors" has become established in biochemistry, where it refers to molecules with highly specific affinities for substances that bind to them, such as hormones. It is thought, for example, that estrogen molecules bind to estrogen receptors in the cell membranes of their target cells, from where the receptors carry them to the nucleus where they can interact with genetic material.

Reciprocal altruism Altruistic behavior between nonrelated individuals, conducted according to the rule that an individual helped on one occasion of need will likewise aid the other when the need for help is reversed. For this behavior to be selected over a purely "selfish" strategy, the benefit of getting help when it is needed, compared to the cost of giving help when it is called for, must be greater, in terms of breeding success, than that of the selfish alternative (see FITNESS). Because an altruistic system can be exploited by individuals that accept help but never return it, who enjoy its benefits without incurring its costs, it is unlikely to evolve unless the "spongers" can be combatted in some way. The conditions would include a group small enough that the same individuals meet frequently and recognize one another individually (see INDIVIDUAL

RECOGNITION), and sufficient cognitive capacity to distinguish and remember who helped and who did not. Such conditions are rare, and so far the only well-substantiated cases of direct reciprocal altruism are from primates and people. Some cases of cooperative breeding in birds have been construed as forms of indirect reciprocity. Speculation about the evolution of human intelligence has included consideration of the social and cognitive sophistication required for reciprocal altruism as a possible factor. The concept was introduced to sociobiology by Trivers (1971).

Recognition Retentive discrimination. The ability to differentiate between two categories of stimuli, but as properties of objects, places, or individuals rather than mere sensory discrimination (see PERCEPTION). In most cases recognition is a result of learning (as in individual recognition), but there are exceptions, such as many cases of species recognition and some of kin recognition.

Recursiveness In linguistics the possibility of repeating a pattern without limit in the construction of sentences. For example, "Tom went" and "Dick went" can be combined into "Tom and Dick went," and this can be repeated for any number of Harrys. Recursiveness allows for such constructions as the multiple embedding of subordinate clauses. Some recent studies of sequence structure in animal communication patterns have revealed the presence of recursiveness (for example, repeated syllable insertions in birdsong), along with recombination of elements—the same elements differently ordered or combined in different sequences.

Redirection Action deflected from the object arousing it and directed at some neutral (substitute) object. By far the commonest form is redirected attack in agonistic situations. For example, if an animal is threatened by a higher-ranking member of its group, it may counter by starting to attack its antagonist but then veer off to land the blow on a lower-ranking "innocent" individual or some object such as a stone or a clump of grass. The grass-pulling of gulls is a typical example. Redirection differs from displacement activity in that the latter is a switch in the kind of behavior instead of a switch in target. Both are found in conflict situations.

246

In Freudian terminology a somewhat analogous phenomenon in human interaction is referred to as displaced aggression or displacement, which should not be confused with displacement activity in the ethological sense, which has its closest Freudian parallel in the concept of sublimation.

Reductionism   Explanation of the molar in terms of the molecular; accounting for surface appearances by showing them to be consequences of an underlying structure or process of a more fundamental order of regularity. In behavior study reductionism usually takes the form of showing how behavior is consequent on physiological mechanisms, which in turn can be understood in terms of their physics and chemistry. Reductionism is typically atomistic and deterministic.

Redundancy   In information theory redundancy exists when more signs are used to encode a message than would be sufficient to do so. It is deliberately included by the communications engineer when noise is likely to disturb signals in transmission. Redundancy is measured as the ratio of the difference between the most efficient coding and the observed coding to the most efficient coding. In a language such as English redundancy enables us to read a text full of typographical errors: differences in the probabilities of letter sequences and of word combinations, syntactic constraint, and semantic connection combine to provide clues to what the words are supposed to be in most of the garbled cases. Most animal communication is repetitive and structured in ways that make it highly redundant. The genetic code is redundant in that many of the amino acids to which it refers are represented by more than one triplet sequence of bases.

Redundancy also exists in the central nervous system. Vertebrate brains, for example, have multiple mapping of sensory input, and the brain's considerable capacity for recovery from neural damage indicates a large degree of equipotentiality.

Reflex   Automatic response of an effector organ to a triggering sensory stimulus. Reflexes have especially rigid stimulus-response relationships, and for many a fixed neural pathway can be specified. The human knee jerk, for example, involves only two nerves—one afferent and one efferent—and the single synapse between them in the lower spinal cord. Hence it is referred to as a monosynaptic reflex. Reflexes established prior to learning are referred to as unconditioned reflexes (for example, pupillary reflex, knee jerk, salivary reflex, clasp reflex). Such reflex responses can be associated with stimuli other than the originally effective stimuli by conditioning, in which case they become conditioned reflexes. The process of establishing conditioned reflexes is sometimes called Pavlovian conditioning, after the Russian psychologist who is generally credited with discovering it.

Reflex arc The anatomical basis of a reflex, that is, the neural pathway that when activated effects a reflex response. A reflex arc begins with a receptor, proceeds via an afferent path to the central nervous system and from there runs via an efferent connection to an effector organ. The simplest reflex arcs consist of an afferent neuron, an efferent neuron, and the single synapse between them (monosynaptic reflex). If the reflex chain consists of three neurons it is called a disynaptic reflex, and the neuron intervening between the afferent and efferent neurons is called an interneuron or internuncial neuron.

Reinforcement Strengthening of a response by making presentation of some reward contingent upon its performance, that is, increasing the probability or frequency of recurrence of the response through reward. There is a distinction between positive and negative reinforcement. In positive reinforcement a "genuine" reward is provided, such as food for a hungry animal or water for a thirsty one, or even electrical stimulation of a pleasure center in the brain; negative reinforcement consists in the withholding or cessation of aversive stimulation. (However, some authors take negative reinforcement to mean the reverse: the infliction of punishment to suppress a response it is associated with, a procedure that many learning theorists consider to have limited effectiveness; but see AVOIDANCE CONDITIONING). In some circumstances something associated with a reward as a sign of it or a means of obtaining it may serve as a reinforcer in that an animal will begin to work for it as vigorously as for the primary reinforcer. For example, monkeys will readily learn to work for tokens that they can later use to get candy from a food dispenser. This process is called secondary reinforcement. When

248

reinforcement is discontinued, the learned behavior associated with it lapses and is said to undergo extinction. The rate of extinction varies according to the previous schedule of reinforcement, being more rapid for fixed than for variable schedules and taking much longer for negative than for positive reinforcement.

Reinforcement plays a decisive role in many forms of learning (see CONDITIONING). Indeed, one proposed definition is that reinforcement is whatever has to be added to contiguity of stimulus and response to effect their association. However, some psychologists maintain that certain forms of learning, such as perceptual learning, do not require reinforcement, arguing that the bare experience of stimulus conjunctions can be registered and remembered and drawn on when it is relevant to goal seeking (see LATENT LEARNING). In classical or respondent conditioning, the unconditioned stimulus is the reinforcer; in instrumental or operant conditioning, reinforcement is consequent on response. In drive reduction theories of learning, reinforcement is postulated to be diminution of a state of need (reduction of a drive).

Reinforcement schedule In a learning experiment, the distribution of reward with respect to occurrence of responses. On a continuous reinforcement schedule the animal is rewarded for every correct response; on a fixed interval schedule it is rewarded when making a correct response after a set period of time; on a variable interval schedule it is rewarded for a correct response after intervals that vary randomly within certain limits; on a fixed ratio schedule the animal is rewarded after it has made a certain set number of responses; and on a variable ratio schedule the rewards are distributed randomly with respect to how many responses the animal has made. Different schedules give differing results, for example in how the animal distributes its activity in time and in the resistance to extinction.

Relationship In ethological writing this term is applied to animals that have established a spatial or social orientation, manifested in consistent and recurring interaction over a sustained period of time. A special form of social orientation is bonding. The existence of a relationship is often difficult to establish methodologically, requiring detailed and precise description and quantification of the frequency, kind, quality, temporal patterning, and directionality of the interactions.

Releaser   A general term for any bodily feature or behavior pattern that elicits a specific response from another animal (sex partner, parent or offspring, fellow group member, or symbiotic partner). Releasers include color and form characteristics, vocalizations, odors (such as sex attractants), and various patterns of movement and posture (courtship and threat movements) functioning as communication behavior.

The terms "releaser" and "sign stimulus" are frequently used interchangeably. When a distinction is drawn, it is usually on the question of whether the stimulating feature has been shaped by evolution to serve its signal function: calling something a releaser implies that this is so; calling it a sign stimulus does not. Thus elicitation of response is the main function of a releaser, and to this end it is specially adapted. With a sign stimulus, on the other hand, the influence on behavior is a side effect: a tree whose colored branches make it easier for birds to recognize it as a nesting place, has not evolved for that reason. In the case of predator-prey relations, conspicuous features that could aid either one in detecting the other tend to be selected against. In other words, a releaser has functional significance for both the sender and the receiver of the signal; the functional significance of a sign stimulus is unilaterally on the side of the receiver.

Most releasers pertain to intraspecific contexts, but some relationships between different species (as in symbiosis) necessitate reciprocal communication. The effective stimuli in such cases are referred to as interspecific releasers. An example is the alarm calls of many songbirds that other species also understand. In contradistinction to interspecies releasers, the stimuli serving intraspecies communication are sometimes referred to as social releasers.

Releasing factor   Also sometimes called releasing hormone. A general term for substances produced by cells of the hypothalamus and conveyed from there to the anterior lobe of the pituitary gland via a network of capillary blood vessels called the hypothalamohypophysial portal system. In the pituitary the releasing factors influence the secretion of the pituitary hormones, which in turn regulate the activity of endocrine glands (thyroid, adrenal cortex, gonads), and so may have behavioral effects. However, in the case of the prolactin secretion of the pituitary, the controlling hypothalamic factor is inhibitory; the hormone is released only when secretion of this factor is discontinued. It is therefore called the prolactin-release-inhibiting factor.

Releasing mechanism (RM)   Stimulus filtering mechanism. A general term for all sensory and neural components that contribute to the filtering of incoming stimuli and hence to determining that only the "relevant" stimuli (see SIGN STIMULUS) trigger behavior at any time; these stimuli signify that in this biological situation the corresponding behavior can have the functionally appropriate effect (for example, stimuli emanating from prey selectively release the corresponding prey-catching action of a predator). This filtering is usually not a unitary process but a series of regulatory steps.

If an animal reacts to a stimulus pattern appropriately without previous experience and thus with no possibility of having learned it, an ethologist is likely to use the term "innate releasing mechanism" (IRM). This term signifies that the connection between a stimulus and its associated response is independent of experience. However, as ethologists have become more aware of the subtle and indirect ways in which experience can affect behavioral development, they have used the term less freely. In the contrasting case the characteristics of the releasing stimulus are learned, and this is referred to as an "acquired releasing mechanism" (ARM). A third, intermediate form is an IRM supplemented, completed, or modified by experience. Such a releasing mechanism is undoubtedly dependent on innate foundations, but in the course of ontogeny, is added to, refined, or altered in some way through learning. The existence of an inborn basis is apparent when the behavior pattern can be released by the IRM alone at an early stage of development.

An example of all three forms of releasing mechanism is furnished by following-response imprinting. Many young precocial birds have an IRM for the following response, which answers to a very few simple visual and/or auditory stimuli. This IRM suffices to release the first following response. During the first following, however, the releasing mechanism is supplemented through learning of a set of recognizable parental features such as body color, and subsequently

251

the following response can be elicited only by the now more refined product of the experience. In contrast, if the young animal is raised by a member of another species, with characteristics that do not correspond to the IRM, it imprints onto these "wrong" characteristics, and thereafter preferentially follows the alien foster species (see ERRONEOUS IMPRINTING). In these animals the groundwork of the IRM fails to participate in the eliciting of the following response; the releasing mechanism is a pure ARM.

IRMs are more common in invertebrate species than in vertebrates, where they tend to be most prevalent in early stages of behavioral development.

Relict   Evolutionary vestige. An organ or behavior pattern that has become functionless in the course of evolutionary history yet persists, though usually in fragmentary or greatly reduced form. Perhaps the best-known examples of relict structures are the vestiges of a pelvic girdle in many snakes and whales. Relicts frequently appear only transiently during embryonic and early juvenile development (for example, gill slits in various terrestrial vertebrates). They can give evidence of phylogenetic affinities and hence contribute to zoological systematics.

Many behavioral relicts continue to be performed even though the part of the body involved has changed so much in the course of evolution that it is impossible for the movements to serve their original function. For example, many macaques perform balancing movements with their tails, even though these are now so short and stumplike as to be quite useless as an aid to balancing. Many pigeons and tree-nesting rails, which are no longer able to retrieve eggs dislodged from the nest because of the placement of the nest, show the egg-rolling movement typical of their ground-nesting ancestors when given an adequate platform.

Relicts often appear in the course of recapitulation. A behavior pattern that was viewed in this light in the early ethological literature is the behind-the-head scratching of many bird species (see HEAD SCRATCHING), but its interpretation is now a matter of some dispute.

Relief ceremony   See Nest relief

Reproduction In biology, the propagation of new individuals. It is one of the most important attributes of life, and its continuation a necessity for species survival. It may be either asexual or sexual. Behavior patterns play a crucial role in the sexual reproduction of many animals (see COPULATION; FERTILIZATION).

Reproductive behavior The general term for any behavior having to do with breeding. The term is used differently by different authors. Although many scientists restrict its reference to the behavior patterns of pair formation and mating, others include in addition territorial behavior, nest building, brooding, parental care, and other categories having some connection with reproduction.

Reproductive isolation Also called sexual isolation. The establishment of barriers to reproduction between different species or incipient species. It is of particular importance where two or more closely related species occur in the same region (see SYMPATRIC). Because such species are commonly able to interbreed and produce hybrids (as shown, by their interbreeding in captivity), the "danger" of miscegenation is especially great here. Sometimes a certain degree of reproductive restriction can obtain between different populations of a single species, especially when they are adapting to differing ecological conditions.

Reproductive isolation is achieved by so-called isolation mechanisms. These include morphological features (incompatible genitalia, called "mechanical isolation") and physiological factors (breeding cycles phased at different times of the day or year). Behavior plays a predominant role in preventing pair formation across species or group lines (called "ethological isolation"); one of the more important functions of courtship behavior is the maintenance of reproductive boundaries. For this reason the courtship of many species consists of an action chain that decreases the likelihood of misalliance. Also the sexual releasers of closely related sympatric species are often very different and thus unmistakable (see CHARACTER DISPLACEMENT; SPECIES RECOGNITION). The early and persisting fixation on characteristics by means of which the sexual partner is recognized (see SEXUAL IMPRINTING) can likewise contribute to maintenance of reproductive isolation.

Resource-holding potential (RHP)   An index of competitive ability. The term was introduced by Parker (1974) in reference to variation among males in their ability to obtain and defend possession of resources such as territory or a harem. Good predictors of resource-holding potential are size, physical strength, quality of an animal's advertisement display, and "badges of status." Thus a bigger animal can usually defeat a smaller one in a fight. Display features are often correlated with size and hence with RHP. Thus the deeper a toad's croak the bigger he is, because size determines how deep a note he can produce. Similarly a red deer stag's roaring tempo is indicative of his size and age and hence his RHP. RHP plays a role in both forms of sexual selection: in intrasexual competition it may affect whether a contender will press a challenge; in intersexual competition it may be a factor in mate choice.

Reticular formation   See Arousal

Retrieving   Carrying back. In animal behavior study retrieving most often refers to the recovery of young by rodent mothers: if pups are outside the nest the mother picks them up one by one in her mouth and carries them back into the nest. She uses the same technique to transport the pups from one place to another. The term is also sometimes applied to the egg rolling of ground-nesting birds.

Retrojektion   A German term for which there is no precise English equivalent, sometimes used to refer to an animal's using its own body, or parts of it, as a substitute object. Especially common are the sucking movements of young mammals: deprived of access to a mother's nipples, as when hand reared, such young suck on paws or fingers or other parts of their bodies. In social isolation experiments young primates often develop the habit of hugging their own bodies with their arms, thus using themselves as substitutes for the mother's body (see CLINGING YOUNG; SEPARATION SYNDROME). Animals kept alone for long periods may also show sexual retrojektion, handling their own genitalia to derive sexual gratification.

Reward   See Reinforcement

Rhythmicity   See Periodicity

254

Ritualization The evolutionary transformation of nondisplay behavior into display behavior; evolutionary change in the direction of
enhancement of signal function; the process effecting such change.
The concept is best exemplified by the displays of visual communication, less well by communication in other modalities since it is
rarely possible to conjecture what the evolutionary antecedents of
vocal or chemical or electrical signals might have been. Even so,
enhancement of signal function involves some principles common
to all—the specifications for efficacy in a communication system.
Hence the evolutionary changes typically are those that increase the
probability of signal detection and reduce ambiguity: increase of
conspicuousness by simplification and "exaggeration" of form, repetition (usually rhythmical), emphasis of particular components, slowing down or speeding up of performance, addition of morphological
support such as coloration, and stereotypy. In contrast to their unritualized antecedents, ritualized behavior patterns typically show
considerable constancy in the vigor (see TYPICAL INTENSITY) and
rapidity with which they are performed. Thus the pecking movements of woodpeckers, which are adjusted to circumstances when
used to obtain food, have a uniform, species-specific rhythm during
drumming. Ritualized patterns are most common in the contexts of
courtship and agonistic behavior.

In most of the cases studied, ritualized movements have been
traced to evolutionary origins in intention movements, ambivalent
behavior, displacement activities, or redirection, augmented by additional secondary signal characteristics. An example is the displacement bill wiping of finches, derivatives of which occur in the courtship of many species. In ritualized form it is performed at a specific
point in the action chain of the courtship sequence, and always in
the same stereotyped manner, often markedly slower than the unritualized comfort movement. In some species the bill no longer
touches the substrate, and the whole movement resembles a bow,
whose origin in bull-wiping would hardly suggest itself were it not
for comparison with other species showing intermediate forms. In
such comparison the ethologist seeks to identify behavioral homologies. The change of function, and presumably of causation, accompanying ritualization has been referred to as emancipation.

Originally the concept of ritualization had only a phylogenetic
connotation. However, an analogous "formalization" can occur in

ontogeny, as when highly variable behavior in young animals gives rise to rigidly stereotyped patterns (see FORM CONSTANCY) in adults; in songbirds, for example, subsong is transformed into definitive song. This is sometimes referred to as ontogenetic ritualization.

Ritualized fighting In comparison to injurious fighting, a relatively harmless form of combat, which proceeds in accordance with set codes or routines. It is typical of intraspecific aggressive interaction, in which it serves, for example, to repel a rival (from a territory, a sex partner, or a food source), or to determine position in a rank order. In species that possess lethal "weapons" such as horns, hooves, or fangs, these are, as a rule, either "kept in the sheath" or used in such a way as to avoid causing injury. Most ritualized fighting is preceded by intensive threat behavior.

Role A term used mainly in the mammalian literature for a position occupied by particular individuals within a group, entailing assignment of activities with a specific social function, for example guarding and leadership (see DIVISION OF LABOR). Primate groups are especially noted for having roles. Sometimes specific roles are restricted to one sex or to certain age classes. Thus in many species the role of leader is always taken by a male, while in others it is occupied by a female. Sometimes there is a tie between the exercise of a certain role and position within the rank order, even though rank and role are conceptually independent of one another. In contrast to other forms of polyethism, the distribution of roles within a group can change in the course of time.

Round dance See Dance language

r-selection Selection resulting from ecological conditions favoring rapid increase in population size. Species adapted to exploit a transient food supply typically multiply their numbers at a high rate whenever opportunity allows. The individuals of such species typically are relatively small, quick to reach reproductive maturity, and short lived; they provide a minimum of parental investment to each of their many offspring, and their social systems lack such stabilizing features as pair bonding and parental care. Compare K-SELECTION.

Rut   Originally this term referred to the annual reproductive activity of male deer, including the period when the activity takes place, the physiological and motivational state of sexual arousal, and the behavior, called rutting behavior. By extension it is now frequently used in the same senses for mammalian sexual behavior in general, especially when its periodicity is marked. Compare HEAT.

Rutting behavior   See Rut

Satellite male   Peripheral male. Initially this term was applied to sexually mature male ruffs (*Philomachus pugnax*) that lack territories of their own and make do by occupying locations within or at the edges of other males' lek areas. Subsequently such males were reported for numerous other species, representing all vertebrate classes and a few invertebrates (grasshoppers). The details vary from species to species, for example, in the degree of tolerance shown by the territory holders and the persistence with which the satellite males encroach. Territory holders may benefit from the aid of satellite males in territorial defense. For satellite males there may be various benefits, such as access to desirable natural resources and increased probability of acquiring the territory later. In many species, opportunity exists for "stolen" copulations with the territory holders' females (see KLEPTOGAMY), which contributes to fitness if fertilization is achieved. The occurrence of two different male strategies within one species or population exemplifies plastic polyethism.

Scent marking   The use of an odorous substance for marking a territory, a social partner, or the animal's own body (see MARKING

BEHAVIOR). Scent marking is most prevalent in mammals, which generally have superior olfactory capacities (see MACROSOMATIC ANIMALS), but it occurs sporadically in other animals also (many bees, desert woodlice). The kind of substance used in scent marking varies. Canine and feline carnivores, lemurs, and many rodents mark their territories with urine; hippopotamuses spatter feces; various marsupials and rodents use saliva; mustelid carnivores, ungulates, and bumblebees use pheromone secretions produced in special glands (so-called scent glands). Glandular secretions often are used for marking social companions, but in some species urine is used instead (SEE URINE SPRAYING). For the marking of conspecifics the term "allomarking" is gaining currency; chemical self-marking is correspondingly referred to as "automarking," at least in some of the primate literature. Besides labeling territories and social partners, odorous substances sometimes serve other functions. Mice, rats, and many fish discharge alarm substances in the presence of danger (see WARNING BEHAVIOR); and many ants leave odor trails when foraging, which allow them to find their way back to the nest or guide others to where they have been. Similarly the nocturnal prosimian slow loris uses urine trail as a guide between places.

School See Flock

Search image Searching image. Restriction of an animal's interest to a single class of object, presumably as a consequence of a temporary setting of its attention, especially with regard to visual perception. For example, a foraging bird is often seen to concentrate on a single kind of food item. Indeed, the concept of search image has been used almost solely to refer to such selective foraging. So long as the type of food is plentiful, so that search for it is frequently rewarded, an animal will continue to concentrate selectively on the characteristics identifying such food, and the search image will persist. But if the supply starts to dwindle, so that return for effort drops below some threshold, the animal will turn its attention elsewhere, sampling other kinds of available food, until the type found most easily becomes instated as a new search image. A search image can be thought of as a temporary releasing mechanism requiring continual reinforcement to be sustained. For certain kinds of animal and food supply combinations, search image feeding is the optimal strategy for the animal to follow in utilizing the available resources. For example, it conduces to the quick finding of cryptic, difficult-to-recognize prey.

Secondary sex characteristics   See Sex characteristics

Secretion   This term refers both to the production of a physiologically active substance by a cell or gland and to the substance so produced. As far as behavior is concerned, two kinds of secretions are especially important: hormones and pheromones. Secretion by nerve cells is referred to as neurosecretion if the site of action of the substance is an appreciable distance from the site of discharge. (Without this qualification the large majority of nerve cells would have to be regarded as neurosecretory, since most synaptic transmission is effected by discharge of a transmitter substance across the narrow space, or synaptic cleft, between presynaptic and postsynaptic membranes.)

Selection   As a consequence of genotypically based variation in phenotypes, some individuals in a population may leave more offspring, and hence more replications of their genes, than others. The resulting change in gene frequencies in the population is selection in the biological sense. However, the term is applied also to the factors that favor some variants or combinations of traits over others. When a word is thus used for both the causes and their effects, there is the likelihood of confusion and circular argument, so care should be exercised in using the term. Fortunately, the context usually makes apparent which sense is intended.

Two kinds of selection, natural and sexual, have been distinguished, although this distinction is viewed differently by different evolutionary biologists.

Darwin (1859) initially proposed natural selection as the paramount evolutionary agency: selection pressures from outside the population, such as food supply and predation, affect an organism's chances of survival and reproduction (fitness); that is, those individuals with the highest probabilities of survival and reproductive success are the ones that have the most fitting characteristics and hence are best adapted through their heredity to their environmental conditions. Selection thus leads to an increase in the frequency of certain genes in a population and a decrease in the frequency of other, comparatively less favorable genes. It consequently complements mutation as a main factor of evolution. The intensity of natural selection is greater under changing environmental conditions than in constant conditions. Furthermore, it regulates the number of offspring a species produces. Species with high mortality rates because of predation or other environmental factors must propagate profusely to offset large losses, for the majority of individuals will fall victim to selection pressure even before reaching sexual maturity. In the sociobiological literature such a situation is described as subject to r-selection. In contrast, K-selection applies to species, mostly large and long-lived, that produce few offspring, each of which has a relatively high probability of surviving to maturity. As a rule K-selected species invest more in parental care (see PARENTAL INVESTMENT).

Sexual selection, which Darwin distinguished from natural selection in his later writings (Darwin 1871), involves the effect on individual reproductive success of competition among conspecifics. It takes two forms: intrasexual selection and intersexual or epigamic selection. Intrasexual selection refers to direct contention between two or more individuals of one sex for reproductive access to or possession of members of the other, usually males competing for females but occasionally the reverse, as in some harem-forming mammals, where several females may come into estrus simultaneously and vie for the attention of the single male. Intersexual selection comes about through the individual preferences of members of one sex, which favor particular conspecifics of the other sex during pair formation or transient mating (see ASSORTATIVE MATING). It can thus be described as a result of mate choice. In the majority of species so far studied, it is the female who makes the final decision, so the male is much more subject to this kind of selection. Sexual selection has an especially strong influence on polygamous species, in which many males live their whole lives without once succeeding in mating.

Selection pressure from within a population may not be confined to sexual competition; it is inherent in social behavior in general, for unless an individual develops social competence, for example in communication, it will not succeed reproductively and its genes will be selected against. It has therefore been suggested that the distinction between selection pressures from outside the population and those from within might be expressed as a distinction between extrinsic selection pressure and intrinsic selection pressure.

Consideration of how helping or hindering the reproductive success of a relative can affect an individual's genetic replication, in addition to its own reproductive effort, has led to kin selection theory and the concept of inclusive fitness.

Selective breeding   Artificial selection. Control exercised by people over the breeding of animals in their care. By choosing combinations of specific hereditary traits and allowing only individuals with these combinations to mate, breeders can create strains from which undesirable characteristics are "bred out" and desirable characteristics enhanced. In this way the domestic breeds of dogs, cattle, sheep, and horses have been produced. In addition to bodily characteristics such as coat color or—as with sheep and cattle—horn shape, selective breeding can affect physiological characteristics, such as milk or egg production, and behavioral characteristics. Examples of the latter are the increased aggressiveness of cultivated Siamese fighting fish and the exaggeration of specific movements in many races of domestic pigeons (tumblers). Many of the characteristics selected for in these cases would be liabilities under natural conditions. Selective breeding thus often goes in directions other than those that natural selection would follow. Characteristics that distinguish domestic strains from their wild ancestral types are referred to as domestication traits.

Selective breeding is also used in scientific experiments in genetics, including behavioral genetics.

Self-marking   An animal's applying an odorous substance to its own body for purposes of communication. It occurs, for example, in hedgehogs, musk oxen, and, most conspicuously, various primates. In some species of monkeys the females dress their tails with urine at the time of estrus. The question of the biological significance of self-marking is still more or less open, and no doubt the answers differ from species to species.

Self-marking, sometimes called automarking, can be contrasted with allomarking, in which an animal applies an odorous substance to a social partner. See SCENT MARKING.

Semantics   See Communication theory

Semiotics   Alternatively, semiotic (by analogy with "logic"). The study of signs and signification. It embraces parts of linguistics, the advertising industry, and the study of communication. The branch of semiotics having to do with animal signaling is referred to as zoosemiotics.

Sensation   See Perception

Sense cell   See Receptor

Sense modality   See Modality

Sense organ   An organ specialized for the reception of sensory stimulation. An arrangement of sense cells (see RECEPTOR), nerve cells, and supporting tissue, including protective and structurally supporting cells, and noncellular material, and pigment shields.

Sensitive period   Sensitive phase. A stage of life in which an animal is especially susceptible to certain learning experiences. As a rule this period occurs very early in an individual's ontogenetic development, but particular external influences (social contact or ecological conditions) at later times can produce stable, long-lasting impressions of comparable or greater degree. The timing and duration of the sensitive period for a particular learning process varies considerably between species and even within a species. Especially distinct are the sensitive periods for the different forms of imprinting; in numerous cases of following-response imprinting, the period lasts only a few hours. Susceptibility to deprivation syndrome is also bounded by a sensitive period, usually of longer duration.

   In the earlier imprinting literature (and sometimes even today) the term "critical period," taken from developmental embryology, covered what is now called the sensitive phase or sensitive period. Because "critical period" connoted sharper and more fixed temporal boundaries than have been found by research in many instances, the newer term is now preferred. Precise timing is critical only in the sense that the opportunity for influence passes; the point to be emphasized is that the period is a time of optimal receptivity to specific environmental or social stimulation. Even "sensitive period" suggests a more demarcated span of time than is consistent with the fuzzy margins of the stage of susceptibility found in many cases. Thus "sensitive phase" can be considered the more appropriate term, although "sensitive period" is currently used more.

As a rule the sensitive period for a particular behavior pattern occurs only once during an animal's life. However, there are exceptions, such as the olfactory imprinting of mother goats and sheep on their newborn kids and lambs; although the sensitive period lasts for only the few hours immediately following birth, this kind of imprinting occurs after each birth, and so can occur several times in the life of a female goat or sheep.

Sensitization   See Priming stimulus; Pseudoconditioning

Sensory coding   The transformation of stimulus effects into patterns of neural transmission, for example, the coding of sound wave patterns impinging on the eardrum into volleys of nerve impulses in auditory neurons specifying the frequency. In principle the process is comparable to the way patterns of vocal utterance are transformed into patterns of electrical transmission when one speaks into a telephone.

Sensory deprivation   See Isolation experiment

Sensory modality   See Modality

Sensory nerves   See Afference

Sensory system   The arrangement of receptors, sense organs, and sensory pathways (see AFFERENCE) to the central nervous system and the receiving or projection areas in the central nervous system that serve a particular sensory modality. Thus, in a mammal or bird, the visual system includes the eyes, with receptors located in the retinas, and the optic nerves, which project to the thalamus, where they connect with a thalamocortical pathway delivering the visual input to the primary projection area for vision in the occipital lobe and with a pathway to the optic tectum in the midbrain. Our understanding of how such systems work has been greatly advanced by recording from single units (see NEURON) at various points in the system and selectively stimulating the sensory surface to find the receptive fields of such units.

Separation syndrome   Separation trauma; hospitalism. Abnormal behavioral development resulting from social isolation in early infancy, shown by apathy, restlessness, compulsive movements (see STEREOTYPY), inability to form social attachments, and other deficits in social responsiveness. The syndrome is of concern to both ethology and human psychology, and has been investigated mainly in socially living mammals, especially primates. Compare DEPRIVATION SYNDROME. See also ISOLATION EXPERIMENT.

Sequence analysis   Recording of the order in which behavioral events occur and subsequent analysis to determine whether the order is random or structured. It can be applied to a single animal, as in analysis of preening bouts or to interactions involving two or more animals, as in analysis of courtship or agonistic behavior sequences. Ethologists have used a variety of approaches both in the recording of sequential data and in its quantitative analysis. Among the more commonly employed statistical methods is stochastic analysis, which involves computing the transition probabilities among different sequence combinations and seeing whether they depart from randomness. Results from sequence analysis can suggest or test hypotheses about the control of the temporal patterning of behavior.

A simple version of sequence analysis has been used in attempts to interpret display behavior as derived activity (see RITUALIZATION). For example, Tinbergen (1959) used observations of the behavior that most often preceded and followed a display, along with evidence from form analysis and situation analysis, to arrive at judgments about the evolutionary antecedents of the postural displays of gulls.

Sex advertisement   Sexual solicitation. These and like terms are used with some variation of meaning in the ethological literature. As a rule, sex advertisement covers all behavior and appearance that may contribute to pair formation in species that lack pair bonds, to copulation. It is also used in a narrower sense to refer to the behavior patterns of courtship. In general sex advertisement pertains only to the sex that initiates pair formation or takes the more active role, the male in most cases.

Sex attractant Pheromone whose function is to attract and, in some cases, guide sex partners to the source of the signal. Sex attractants are found especially in insects. In some Lepidoptera, such as the silk moth, and many Hymenoptera, such as honeybees and Pharaoh's ants, the males are attracted and sexually stimulated by a pheromone secreted by the female. A familiar example from mammals is the scent of a bitch in heat, which attracts the male dogs within range of her smell. The olfactory sensitivity is in many cases remarkable. For instance in some Lepidoptera the males can detect and orient themselves toward a female discharging sex attractant at a distance of several kilometers. Sex attractants can play a role also in bringing individuals together in the nuptial swarms of social insects; see NUPTIAL FLIGHT.

Sex characteristics A general term for those characteristics of sexually dimorphic organisms in which the two sexes differ. The reproductive organs in the narrow sense, that is, the gonads with their ducts and accessory glands, are referred to as primary sex characteristics. Secondary sex characteristics are all the additional distinguishing features, such as body size, coloration, or form.

Differences in body size are very marked in many invertebrates and in some fish species. In such cases one sex, as a rule the female, may be many times larger than the other (so-called dwarf males). Considerable size differences occur also in mammals, most typically in harem-forming species, in which competition for females has evidently led to the gradual evolution of increasingly more powerful males. Differences in coloration of the sexes occurs most prominently in birds and fish. On the whole, striking secondary sex characteristics are more often a feature of the male sex, for example, antlers and horns in many ungulates (although in some species both sexes have horns, and both male and female reindeer have antlers), and especially bright coloration in birds and fish. Body size and appendages are permanent secondary sex characteristics, in contrast to those that are confined to the breeding period, such as breeding plumage in many birds (see ADVERTISING DRESS; COLOR CHANGE). In such cases the conspicuous characteristics, which are often

266

shown off by means of specific movements, are usually important releasers.

Sex hormone Sex steroid. Any of the hormones secreted principally by the gonads that regulate the many developments and processes having to do with reproduction, such as formation of secondary sex characteristics, which generally have some releasing function pertaining to reproductive behavior patterns (for example, courtship, nest building, parental care). In vertebrates the male sex hormone is referred to as androgen, and the female sex hormones are estrogen and progestin.

Sexual dimorphism Sex dimorphism. Differences in appearance (size, shape, coloration), physiology, and behavior, between males and females. In both sexes differentiating characteristics are referred to as sex characteristics. Sexual dimorphism in body color (sex dichromatism) often represents an adaptation to the environment, as in the case of the inconspicuous appearance ("plain clothes") of the sex assigned to parental care.

Sexual dimorphism may also be of significance in social contexts, as when particular characteristics of one sex, for example vocalizations or color markings, act as a releaser on the other. As a rule sexual dimorphism is more strongly developed in polygamous species. Species in which the sexes look and behave similarly are referred to as monomorphic (see MONOMORPHISM).

Sexual imprinting A learning process occurring in a young animal that determines what its preferences will be in pair formation. Originally it was assumed that only species-specific, supra-individual characteristics were learned through sexual imprinting and that the learning process served chiefly for species recognition. However, more recent findings indicate that, just as with filial imprinting, sexual imprinting can lead to discrimination of individual characteristics and so contribute to kin recognition and hence assortative mating.

Sexual imprinting presents a typical case of imprinting in that, for a large number of animal species, a yearlong or even lifelong irreversibility can be demonstrated. If a young animal of a sexually imprintable species is raised by foster parents of another species, it will show the effects of erroneous imprinting.

Sexual isolation   See Reproductive isolation

Sexual selection   A form of selection pressure exerted from within a population through social competition for reproductive opportunity. Darwin, who introduced the concept to biology, distinguished two kinds: Intrasexual selection involves competition between the members of one sex—males in the large majority of cases—for possession of the means of sexual access to the other sex, such as territory or a harem. Intersexual selection (epigamic selection, mate choice) refers to the choice exercised by members of one sex, females in most cases, in taking sex partners. Darwin believed that sexual selection often works in a direction contrary to that of natural selection; for example, an animal's conspicuous form or behavior that might attract a sex partner could also increase its chances of being spotted by a predator. As instances of the power of sexual selection, he pointed to the unwieldy antlers of the extinct Irish elk and the fantastic tail of the peacock. See SELECTION.

Shaking to death   Vigorous side-to-side head shaking with prey held in the jaws. It occurs in many carnivores (canids, mustelids, viverrids) and a few other mammals (Tasmanian devil, hedgehog). Different views have been expressed on the evolutionary derivation of the action. Its origin probably lies in the arrhythmic and asymmetrical head-tossing movements shown when carnivores seize small prey and then immediately fling them away. In the next stage the prey is shaken back and forth several times before being let go. If the predator loses its grip on the prey, it can bite again to get a more secure hold, without the prey's escaping. The most highly developed form is the true shaking to death used by canids, in which the shaking movement kills the prey by suffocation or neck dislocation. In the young of such canids the shaking-to-death movement is a frequent component of play behavior.

Shaping   The use of operant conditioning techniques to get an animal to perform movement patterns not originally in its repertoire. By reinforcing any movement tending in the direction of the desired pattern, a trainer can get an animal to come progressively close to it and eventually complete the mastery of it (see REINFORCEMENT). In this way circus trainers are able to produce bicycle-riding bears, ball-balancing seals, waltzing elephants, and so forth. Skinner (1963), who has used shaping for behavioral research purposes, trained a pair of pigeons to play ping-pong. An animal's initial repertoire of movements has to include acts for the shaping to start on, and it has been found that if the direction of the shaping violates the "natural" connections of a movement to its function, the animal may resist the training so much as to make it ineffective (see LEARNING DIS-POSITION).

Short-term memory   See Memory

Shyness   See Avoidance conditioning

Sibicide, siblicide   See Fratricide; Infanticide

Signal   As a verb, to send a signal. As a noun, whatever passes from a sender to a receiver via a channel in an instance of communication. For animal communication the channel can be optical, in which case the signal is light emitted or reflected from the sender; acoustic, in which case the signal is a sound of some sort; chemical, the signal a pheromone; tactile, the signal a touch; or electrical, the signal a pulse of electricity generated by the sender. However, when we talk of an animal's signals we often refer to the actions it takes to communicate. We refer to postures or movements as signals, even though it is the light they reflect that is registered by the recipient. Ethologists also use some freedom in what they take to be the boundaries of signals, sometimes regarding the notes, sometimes the phrases they constitute, as the signals. Where signaling is continuous, as in olfactory communication, there are no boundaries and the signal is simply the scent. To be effective, a signal must have a certain degree of improbability, must stand out from the background of random events or surrounding flux; that is, the signal-to-noise ratio must be sufficient for the recipient to detect the signal. A consequence of this is that there can be adaptive correlation between the characteristics of animal signals and the characteristics of the habitats in which they are used (see MELOTOPE).

Sign language See Language

Sign stimulus Key stimulus. An external stimulus, usually either a single simple feature or a compound of a few simple features that provide only a small fraction of the total sensory input from a situation to an animal, but to which the animal's specific reaction pattern is tuned, so that the stimulus selectively elicits this pattern. If the stimulus is a social signal, evolved specifically for communication (for example, certain scent substances, vocalizations, or patterns of coloration), it is referred to as a releaser. Besides their releasing function, sign stimuli can influence the orientation of a behavior pattern and affect motivational state (as a priming stimulus, say).

The selective responsiveness associated with sign stimuli is an extreme form of stimulus filtering. German ethologists favor the term "key stimulus," partly because the external stimulus suggests a key that unlocks the "gate" of the relevant releasing mechanism.

Sign vehicle The effective part of a signal; the feature that affects the behavior of the recipient. For example, experiments on the pecking response of herring gull chicks to the signal presented by a parent's bill showed that the color of the spot on the lower mandible, and its contrast with the rest of the bill, affect the strength of the response, but that the color of the rest of the bill does not. Accordingly, the color and contrast of the spot contribute to the sign vehicle, and the color of the rest of the bill does not. As this example shows, in many behavioral contexts, the sign vehicle is the same as the sign stimulus. However, sign vehicles also include stimuli that act over an extended period to produce physiological change, such as the components of a male dove's coo that induce ovarian maturation in a female. The term derives from the communication theory of Morris (1924).

Silverback male See Status signal

Simultaneous choice test See Choice test

Site tenacity See Fidelity to place

Situation analysis Context analysis. In seeking evidence upon which to build arguments about the evolutionary derivation of displays (see BEHAVIORAL PHYLOGENY; RITUALIZATION), ethologists have noted the circumstances in which the displays are performed in the hope of finding clues to their unritualized roots. For example, Tinbergen (1959) observed that the "upright" posture of gulls occurs most often in situations where a bird is simultaneously stimulated to both attack and flee. This, added to other evidence from form analysis and sequence analysis, supported his interpretation of this display as a composite derived from intention movements of attacking and fleeing.

Skinner box Puzzle box. A device originating with Thorndike (1911) but named for Skinner (1938), who refined it and whose use of it in studies of operant conditioning is perhaps the best-known and most influential work in the psychology of learning. The device consists of a receptacle for the experimental animal, a "manipulandum" (a disc at which a pigeon can peck or a lever that a mouse or rat can press), and a "reward dispenser" (for delivering food or water as reinforcement). There may also be some means of presenting the stimuli. By coupling manipulandum and dispenser in various ways, an experimenter can operantly (instrumentally) condition the animal and investigate how such independent variables as the reinforcement schedule, magnitude of reinforcement, reinforcement latency, and intervals between trials affect such dependent variables as learning rate, resistance to extinction, and temporal distribution of activity.

Sleep A condition of quiescence characterized by immobility and minimal responsiveness to stimulation, usually occuring for periods that follow a circadian rhythm. It is a widespread phenomenon, at least among vertebrates, but the details vary so much from species to species that a comprehensive definition covering all cases remains elusive. For example, the amount of time spent in sleep daily ranges from twenty hours for the two-toed sloth to less than an hour for shrews. A sleeping animal's lack of responsiveness to stimulation may be extreme, as in fish that can be lifted out of the water without struggling when asleep, or may be little different from the waking state, as in many herbivorous mammals subject to predation. Sleep-

271

ing elephants appear oblivous to familar noises but are alerted by even a hushed occurrence of strange sound. Most species confine their sleep to particular times of day, that is, they are diurnal or nocturnal, and the choice often coincides with adaptive considerations. For example, many lizards sleep at night, when temperatures are too low to sustain their metabolism at the level necessary for activity, and many small mammals sleep in the day, when activity would expose them to a high risk of predation. Animals may prepare for sleep by seeking out a suitable and safe place or by constructing a nest of some sort. Such appetitive behavior and the species-characteristic nature of sleep have led some ethologists to the view that it has its own drive or motivational system.

Physiological studies have confirmed that sleep is controlled by its own brain mechanism, at least in higher vertebrates. Electrical brain stimulation and lesioning have located specific sites involved in the regulation of sleep. The sleep state itself has been investigated by EEG recording (see AROUSAL). This has shown that mammals and birds alternate between quiet sleep (QS), during which the brain waves are relatively long and of high amplitude, and active sleep (AS), during which the brain waves are relatively short and of low amplitude. The latter pattern is similar to that of the waking state, and there are other indications of arousal, yet the sleeper is, if anything, harder to waken than at other times. During active sleep there may be periods of rapid eye movements, and this kind of "paradoxical sleep" is sometimes referred to as REM sleep. In humans rapid eye movements have been associated with dreaming, and the same connection is thought likely in other animals in which it occurs.

The functions of sleep are still a matter of debate. Traditionally it has been thought that sleep is necessary to restore energy reserves used up during wakefulness, but this has never been conclusively demonstrated. Another view is that the periods of immobility and reduced responsiveness, especially if spent in places of safety, remove an animal from situations where it is exposed to hazard, such as the effects of environmental stress or predator attention. Sleep also conserves energy. It may be adaptive for an animal to be active only for as long as and when necessary and to spend the rest of the time out of harm's way. The variations in details from species to species make it likely that the functions of sleep are different from one to another.

Sleeping position The body attitude adopted by an animal during sleep. Most animal species use one or more quite characteristic sleep positions, commonly lying on the belly, side, or back. Bats and bat parrots (so called because of their similarity to bats in this respect) sleep hanging head down from a branch or rock. Most birds sleep on one leg and tuck the bill tip firmly behind the folded wings. Ostriches lie with head and neck flat on the ground. Perhaps the oddest sleeping position of all is that of the giraffe: it folds its front legs, bends its long neck to one side, down and back like a hook, placing the top of the head behind the spread-out hind legs on the ground.

Sleep nest A nest used as a shelter in which to sleep rather than for raising young. Such nests are most common in songbirds (wrens, sparrows) and apes. Songbirds often use vacant brood nests, either of their own or other species, as sleep nests, but will seldom lay eggs there. Many species spend the night singly, others in pairs or larger groups in sleep nests. Their biological significance is still uncertain. At least it can be said that heat insulation is not the only function, for tropical species use sleep nests even more commonly than temperate-zone species.

Social In ethology, pertaining to behavior or relationships involving, usually, conspecifics. However, the term is also sometimes applied to cooperative interactions between members of different species, especially if the relationships resemble those between members of the same species.

Species are referred to as social or gregarious if their members live as enduring pairs, as families, or in larger groups in consequence of which social behavior makes up a major proportion of their total activity. In some species the need for social companionship and the dependence of conspecifics on one another are so great that individuals are incapable of surviving alone. In such obligate social species the assemblages often show division of labor, sometimes (as in many of the social insects) with associated specializations of body form and behavioral capacities. In contrast, in some social species single individuals occasionally go off alone briefly or for an extended period. Such an animal is sometimes called a lone wolf. Compare SOLITARY.

Social attraction   The mutual attraction between conspecifics of social species (and sometimes also members of different species, especially in mixed groups). It is the basis for much group formation. In many cases social attraction may depend upon a motivation peculiar to it (see BONDING DRIVE).

Social behavior   Behavior having to do with interactions between conspecifics or between animals of different species where some form of symbiosis exists in which the behavioral communication is similar to that within a species. As this definition indicates, the ethological concept of social behavior is very broad, and the limits of its reference are not sharply demarcated. Indeed, different writers draw them in different places. Most behavioral scientists take social behavior to include aggressive interaction, such as territorial defense, as well as the more amicable patterns of courtship and parental care; others restrict the term to group behavior of the sorts shown by social animals living in enduring associations. In human contexts the term also has some variants of connotation that do not apply in animal behavior; for example, the sense of social implied when someone's behavior is described as antisocial rarely applies to a case other than the human.

Social communication   See Communication

Social deprivation   See Isolation experiment

Social dominance   See Dominance

Social drive   See Bonding drive

Social dynamics   A general term for all the ways in which a species-characteristic social structure is maintained or varied. See SOCIAL SYSTEM; SOCIOBIOLOGY.

Social facilitation   Contagion. The tendency for an action by some members of a social group to be taken up by others and so spread through the group, perhaps to the point that all or most of the members are doing the same thing at once. Thus a satiated hen can be induced by the foraging of her companions to resume pecking at

274

grain. Social facilitation is an important means of achieving synchronization of behavior within a group. It can be decisive where quick reaction is called for, as in taking flight from a predator (see WARNING BEHAVIOR), or where natural resources, such as watering places, are so widely spaced that it takes a group a long time to locate the next one; an individual that was not thirsty and did not conform to the behavior of the group would be more at risk because it would be unable to travel as far as the others before suffering from water depletion.

Social facilitation should not be confounded with imitation. Whereas imitation results in acquistion of a *new* capacity or way of behaving, social facilitation simply promotes an action that the animal already has in its repertoire. In contrast to the learning involved in imitation, social facilitation puts nothing new into memory.

Another distinction to be drawn is that between social facilitation and mood induction. As a rule social facilitation refers to a situation where the same "mood" underlies the behavior, such as foraging, of all the animals involved; the social stimulation merely influences the timing of the behavior. Mood induction, on the other hand, involves an animal's switching to a new mood (for example, from preening to fleeing) because of stimulation from the behavior of others.

Social grooming Allogrooming, allopreening. One animal in a social group cosmetically treating the skin or coat (hair, feathers) of another. It is generally, but not exclusively, directed to parts of the body that the passive partner cannot reach. Each of the different forms of social grooming probably originated as a means by which animals could mutually clean one another and remove ectoparasites, which provide compensation by being edible. In a number of cases it has secondarily acquired additional social significance, which may even have become its main function, as in promoting pair or group bonding. In primate species with rank-order social organization, rank is relevant to who grooms whom. The grooming roles reflect the social relationship and also probably help reinforce or develop the relationship in subtle ways by affecting the dispositions of the animals toward one another.

Birds use the bill for social grooming; primates use their hands, as in the delousing of many monkeys; and other mammals use the tongue or teeth (licking, nibbling). Social grooming may be unilateral, may alternate between the two partners, or may involve both partners simultaneously. In many species, especially in birds and primates, a special soliciting posture is adopted, which presents the body part requiring attention to the partner; in certain cases its appearance is enlarged by the erection of feathers or fur.

Social hierarchy See Rank order

Social inhibition Suppression of a specific action by one individual through the influence of another. It has been observed in many behavorial contexts. For example, appeasement behavior exerts a strong inhibition on the aggressive tendencies of conspecifics (aggressive inhibition, or inhibition of killing: see KILLING BEHAVIOR). In group-living species certain actions of a higher-ranking individual in the group can block the performance of similar behavior by a subordinate. Thus in many mammal groups the leadership role is always taken by one particular individual. When that individual dies, the leadership is assumed by another member of the group, which indicates that the successor was ready for the role but was "hindered" from adopting it by the presence and behavior of the incumbent. Many primates (for example, hamadryas baboons) manifest a special kind of social inhibition in which the dominant male takes over the females of other males.

Among some mammals a form of social inhibition having a physiological effect has been described as psychological castration.

Socialization Acquisition of social capability through the experience of interaction with conspecifics as a consequence of growing up within a group. Such learning may be a prerequisite for later social accomplishment, such as the formation of relationships, bonding behavior, acceptance into a group, and assumption of roles. In many species some forms of social competence require specific social contacts during a certain stage of development, which may be referred to as a socialization phase. Because of its early timing and short duration, such a phase is similar to the sensitive period of imprinting. If, in an isolation experiment, for example, an animal is deprived of the relevant experience during what should be its socialization phase, it will probably be incapable of forming normal social relationships when it grows up. Bypassed socialization phases are therefore the cause of many of the developmental disabilities of the deprivation syndrome.

Socialization phase   See Socialization

Social modification   See Culture

Social parasitism   A form of parasitism that occurs in the societies of social insects. In contrast to "real" parasites, social parasites (silverfish, various beetles, and the ants known as robber ants) do not invade single individuals but inflict themselves on the colony, for example by consuming food reserves (see KLEPTOPARASITISM) or by killing and eating the brood. As with brood parasitism, social parasitism presents a variety of special adaptations, many of which are behavioral in nature (for example, behavioral mimicry).

Social releaser   See Releaser

Social stimulation   Unilateral or reciprocal stimulation between conspecifics. It plays an important part in courtship behavior, where it contributes to the promotion of reproductive readiness, and also in the activity of group-living animals, for which it can bring about synchronization of flight or departure for migration or the start of breeding. See BREEDING COLONY; FRASER DARLING EFFECT; SYNCHRONIZATION.

Social stress   See Stress

Social system   Social structure, social organization. A species' patterns of spatial distribution and group association. It includes the form of association between the sexes; social relationships outside family bonds, for example group size, group stability, and the use of territory or home range; and relationships between different groups. The social system is adapted to the environmental and ecological conditions in which the species lives. Thus species in habitats such as tropical rainforests, where food supply is fairly evenly distributed in space and time, often have small social units (pairs or a few individuals); inhabitants of open terrain, where food and water are concentrated in specific places, tend to form large groups. However, in some groups, such as weaver birds, ungulates, and primates, this relationship is not very marked, and different species present a great variety of adaptations of social structure to the prevailing living conditions. Closely related species occupying different habitats show quite pronounced differences in social systems (as observed, for example, in kangaroo species and in the Old-World monkeys known as guenons); in some cases the social structure varies even within a species (as observed, for example, in the blue gnu, in lions, in various primates, and some iguanid lizards). Investigation of the correlations between ecological conditions and social organization is one of the main concerns of sociobiology.

Social tendency See Bonding drive

Social tonus A state of mild tension that has been described in mammals of group-living species, although it doubtless occurs in other animal groups as well. It results from the continual stimulation of social encounters and may be a prerequisite for the development of site fidelity (see FIDELITY TO PLACE) and the social bonding necessary for group living. Social tonus obtains only within a certain range of population density. Above or below this range, individual and collective well-being is likely to deteriorate (see STRESS).

Society A sizable group of communally living animals in which the social organization includes division of labor and regulation of activity that can be compared to forms of human society. A prerequisite for the evolution of societies is some advanced form of parental care.

The largest and most rigidly ordered animal societies are those of the social insects: termites, ants, and bees. The societies of Hymenoptera are maternal families consisting of one or more sexually mature females (queens) and a large number of sterile females (worker bees), which are sisters of the queen and of one another. The male bees (drones) remain in the society only until the nuptial flight, when they compete to make their sole contribution to the society, namely fertilization of the queen. Termite societies are made up of both sexes equally and so represent a two-parent family. A male (king) stays on with the queen, and the workers are younger, reproductively immature adults of both sexes. In many ants and termites there are castes among the workers. Within the insect societies food is distributed according to need, usually delivered in liquid form from one individual to another via mouth or anus, a mode of food provisioning called trophallaxus. The activity of the society is regulated in numerous ways by pheromones.

The number of individuals in an ant or termite society can reach the millions. (Populations of South American nomadic ants have been estimated at 20 to 22 million!) Altogether, about 8,000 species of social insects have been identified. Such societies stay intact indefinitely, so there is overlap between succeeding generations. This sets such "eusocial" insects apart from the "semisocial" species, such as bumblebees and wasps, in which the societies are much smaller and do not endure long enough for the generations to overlap.

In addition to the insects, some mammal and bird species with social organizations have been described as societies. In these the organization and regulation of role are looser and more flexible than in insect societies, a reflection in part of the more psychogenic social regulation in vertebrates compared with the biogenic social regulation of insects. Most primate species live in such societies. Among birds the closest approaches are the communally nesting weaverbirds and species in which there are helpers at the nest, such as the long-tailed tit (*Aegithalos caudatus*). Among mammals the eusociality of social insects is approached most closely by the mole rats (*Heterocephalus glaber*).

Sociobiology "The systematic study of the biological basis of all social behavior" (Wilson 1975). A newly differentiated branch of science, that focuses on the social behavior of animals, investigating how the social system of a species is correlated with the habitat conditions to which it is adapted. Sociobiology is primarily interested in the selective value of specific social systems; it measures and assesses survival value and examines how the system under investigation adapts the members of the species to their habitat (see STRATEGY). In contrast to ethology, sociobiology is concerned little with the mechanisms that bring about and maintain social systems; it concentrates on the ultimate factors rather than the proximate factors determining behavior. Accordingly, it applies evolutionary biology to the social behavior of organisms. However, it takes little interest in behavioral phylogeny or such problems as the course of ritualization and the criteria of behavioral homology, which preoccupy ethologists. Nevertheless, like ethology, it is to a high degree a comparative science. Before the name "sociobiology" became fashionable, this field was sometimes called socioecology, and part of it is still often referred to as behavioral ecology.

Sociobiology has undertaken for the first time a really comprehensive biological analysis of social structure, while also investigating population biology and population genetics. It has induced ethology, which has traditionally been more interested in species-specific behavior, to attend much more closely to intraspecific variation, especially individual differences. This perspective has led to the elucidation of altruism, which had previously been left unaccounted for in terms of selection theory (see KIN SELECTION THEORY).

Socioecology See Sociobiology

Sociogram A graphic representation of social relationships or interaction frequencies in a group. Individuals are usually represented as labeled boxes or circles connected with arrows to indicate how often each one does what to whom in a sampling of the behavior of a group. Such an information network can display, for example, the dominance interaction in a hierarchically organized group.

Sociometry Quantitative description of the social relationships within a group, the behavior patterns of its members, and the organization of the group. One of its methods consists of measuring the distances between group members and the directions they face relative to one another (see ATTENTION STRUCTURE). Such measurement provides, among other things, evidence of bonding between the animals. So, for example, in a hamadryas baboon troop, the distance between a female and her male during rest periods is about 60 centimeters, while the distances between other members of the group is some 10 meters.

The methods of sociometry were originally developed in human psychology, where they are used in a similar way to ascertain relationships between particular individuals within groups.

Software See Program

Soliciting behavior Movements and postures of a female inviting a male to engage in copulation with her. It constitutes the last part of female courtship, or "preceptive behavior," indicating sexual readiness. In many animal groups special displays are used for sexual solicitation, for example, presentation by female primates of the genital area, which is often strikingly colored. This is referred to as presenting or genital presentation. In the Old World monkeys the display serves additionally as appeasement behavior.

Solitary A designation used in biology for species in which individuals form no enduring groups or pair bonds, but live most of their lives in a solitary state. Males and females commonly occupy separate territories and meet only for mating. Mammalian examples of solitary animals are sloths, orangutans, many felid carnivores, and the field hamster; avian examples are cassowaries, cuckoos, and European robins.

Sollwert A German word that has been taken into the ethological lexicon. The literal translation is "should-be value," and it means the same as "efference copy" in the reafference principle.

Sonagram See Sound spectrogram

Song In ethology and bioacoustics, usually sound production of a length and complexity that distinguishes it from shorter and simpler calls. However, songs and calls overlap so much in both length and complexity that the distinction is often hard to sustain, at least on morphological grounds. Songs are not limited to vocal utterances; they can be produced mechanically in various ways. As a rule the sound sequences constituting songs vary in some way (frequency modulation, amplitude modulation, or duration), but some insects and amphibians have songs consisting of strings of identical sounds, whose distinctive patterns are rhythmic. In the very complex songs of many bird species it is often possible to differentiate phrases separated by pauses. These in turn can be divided into "syllables" separated by shorter pauses, and finally these may be composed of distinguishable individual elements or notes.

Songs occur in mammals, especially among primates (howler monkeys, gibbons), in amphibians (frogs, toads), in many fish and insects (notably crickets, locusts, and cicadas). They are most prominent in birds, and are by no means confined to the songbirds. In most cases only the males sing. However, in some exceptional species the female sings, either solo (bullfinch) or in association with her mate (see DUET SINGING).

The functions of song vary between species. It can advertise a territory to deter intrusion of conspecific rivals (territorial song); initiate or accompany aggressive encounters between rivals (agonistic or threat singing; see APOTREPTIC BEHAVIOR); attract and sexually stimulate opposite-sex partners (soliciting or courtship song); or assist the bonding and mutual adjustment of a pair, or even of members of a whole group. In most cases the same song serves several of these functions, the different effects perhaps being determined by context of occurrence, but in some species different songs are associated with different functions (see SONG REPERTOIRE).

By withholding specific learning opportunities from young songbirds, scientists have been able to learn about the relative contributions of innate and acquired information to the development of song (see ISOLATION EXPERIMENT). Quantitative studies have helped clarify this frequently discussed issue. For species in which the birds must learn their songs and transmit them to the next generation by tradition, a further important function of singing is implied: to serve as a model for the young. Many bird species have developed different songs in different parts of the distribution ranges (see DIALECT). A form of song whose function is still not well understood is subsong.

Song dialect See Dialect

Song flight See Courtship flight

Song imprinting Said to apply in bird species in which the song is learned during a more or less clearly defined sensitive period, after which there is little or no capability of learning new or additional patterns. Song acquisition of this sort thus shows the essential characteristics of other kinds of imprinting (a sensitive period and irreversibility). However, such song often begins to emerge some time after the end of the sensitive period, so the "imprinting" applies more to acquisition of a template, to which the bird later matches its song production, than to production of the song itself. It is therefore more comparable to sexual imprinting than to following-response imprinting (see FILIAL IMPRINTING).

Song repertoire   In the bioacoustic literature, an inventory of the song types or song phrases available to an individual. The term is used mainly of male songbirds. In the majority of songbirds the males have only one song type, which serves for all functions (see SONG). A single song type may vary individually and geographically (see DIALECT). In some species, however, the males have two or more (in the North American chestnut-sided warbler, five or more) different song types, each of which serves its own function. Thus the Cuban grassquit has a short song that it directs at other males, which serves as threat, and a longer song that is repeated more in the presence of females and so probably has sexual significance. Other species have from a few to many different song types (in the American long-billed marsh wren the number is about a hundred, and in the brown thrasher it may be more than a thousand). Sometimes the size of the song repertoire varies between different geographical races; again the long-billed marsh wren is an example.

The full significance of repertoire size is not yet understood and may differ between species. Observations indicate that it may affect the response by individuals of both sexes: having a large number of songs may, among other things, be a status signal, advertising a male's maturity and experience and so having strong attraction for a female; or a large repertoire may enable a male to persuade other males to move on by giving the impression of a greater population density in the region than actually exists (the so-called Beau Geste hypothesis). For both sexes a large repertoire possibly also facilitates individual recognition.

Sonogram   See Sound spectrogram

Sound isolation   See Isolation experiment

Sound production   Phonology. The term "vocalization" is often used as if it were synonymous with behavioral sound production, but strictly speaking it comprises only voiced sounds. Sound production is a major means of communication. As already indicated, a distinction is drawn between mechanically produced sounds and vocal

283

sounds, which derive from the use of the breath. Mechanical sound production is important for many invertebrates, especially insects, in which it takes such forms as striking part of the body on the substrate (knocking sounds, as in death watch beetles), vibrating the wings (buzzing sounds, as in mosquitoes and honeybees), or rubbing parts of the body (antennae, wings, legs) against one another (rasping, chirping, or trilling sounds, as in grasshoppers and crickets). This latter mode is referred to as stridulation. Mechanically produced sounds are also found in some vertebrates. Examples include the sounds some fish make by rubbing parts of their skeleton against each other; the drumming of woodpeckers, produced by rapidly hammering the beak against wood; and the foot stamping of various mammals. A further means of producing mechanical sound, which is found among both vertebrates (some fish) and invertebrates (cicadas), is by oscillation of membranes, as in an earphone.

With sounds produced by respiratory air flow, a distinction is drawn between simple calls and more complex songs, which are composed of different elements. The study of sound production in animals has contributed important knowledge to a variety of branches of behavioral science.

Sound simulation   See Vocal simulation

Sound spectrogram   Sonagram; sonogram. A graphic representation of sounds, such as vocalizations, in the form of a bicoordinate plot. It is produced on a machine originally called a sound spectrograph or Sonagraph (a brand name), which records the graph on a strip of paper. Usually the machine plots the distribution of sound energy with respect to frequency on the ordinate against time on the abscissa, but some machines can also plot amplitude against time and other display options. There are also choices of frequency and time resolutions, speeds, and frequency ranges, which make it possible to display a great many aspects of a sound pattern. With advances in electronic technology, especially the new machines called spectrum analyzers, graphic representation of sounds has become even more flexible and convenient.

The ability to make pictures of sounds and to use them for making measurements and comparisons has been a boon to the ethologist interested in acoustic communication (see BIOACOUSTICS). As a consequence study of communication in this modality is in some respects in advance of study in other modalities, such as work on individual characteristics and individual recognition.

Speciation Species originate when two separately evolving subspecies diverge to the point where the differences between them amount to between-species differences. As with the general difficulty of defining species, so is it often hard to decide when this point has been reached. As a rule speciation involves a period of geographic isolation, during which the subspecies undergo differential adaptation consequent on the differences between their habitats (see ALLOPA-TRIC). Whether speciation can also occur when the subspecies occupy a single region (so-called sympatric speciation) is a question extensively discussed in the literature; indisputable evidence of sympatric speciation is still wanting. It is at least feasible that a group's very strong fixation on a particular part of the habitat could lead to reproductive isolation from other groups of individuals, which could lead to species-generating divergence. This, of course, may also be viewed as allopatric speciation, the spatial separation simply being on a much smaller scale—microgeographical isolation, as it were.

If two allopatric populations resume contact and if interbreeding is still possible, the hybrids may prove less fit than the offspring from matings within each group; in that case selection will result in the erection of barriers to interbreeding. Behavioral mechanisms, such as increased selectivity during pair formation, frequently contribute to the establishment of such reproductive isolation.

Speciation of a different sort is shown in collections of fossil material. Here a single lineage may change over time to such an extent that forms from later strata may be considered as representing a different species from those in earlier strata. The judgments are made on the basis of morphological criteria and breaks in the fossil record. A species is judged as having been transformed into a different species as a consequence of accumulated difference with the passage of time.

The view of speciation just given represents it as a process continuous with the microevolutionary changes within species. This view has recently been challenged by one that describes phylogenetic history as consisting of punctuated equilibria. According to this, the macroevolutionary processes giving rise to species involve a different kind of selection from the individual or genic selection of intraspecies microevolution.

Species The problem of defining this intuitively natural category of classification for all occasions of its use is an old one; it is now generally agreed that the word can have at least slightly different meanings in different contexts. The connotation of the term in current biology—the biological species concept, as it has been called—makes reproductive continuity the criterion: a species consists of all the organisms that draw from the same gene pool, that is, a reproductive population—a group of individuals that, under normal conditions, freely breed with one another. Between individuals of different species, on the other hand, reproductive isolation is the rule. However, the drawing of species lines can present a difficulty in many cases, especially with allopatric forms for which it is not known whether interbreeding would occur if they encountered one another in natural circumstances. (In captivity members of different species often interbreed and produce hybrids.) Assuming that the separate populations came from a single population initially, they might still belong to the same species or already constitute two different species (see SPECIATION).

Where tests for interbreeding are not possible, the taxonomist may have to fall back on the morphological species concept, according to which the judgment of species status depends upon the degrees of difference in features of form. An extension of this concept to biochemical analysis of molecular structure has brought considerable refinement to the demarcation of species. When dealing with living forms, the morphological concept can be thought of as a working substitute for the biological concept, but when dealing with paleontological material, comprising forms isolated in geological time, it is the only option. The criterion of interbreeding is inapplicable also to asexually reproducing organisms, to self-fertilizing creatures such as many cestode flatworms, and to the many plants that hybridize more easily than is consistent with taxonomic convenience. In animals, along with anatomical, physiological, and biochemical characteristics, patterns of behavior can be used for taxonomic purposes.

In biological systematics, closely related species constitute a genus, closely related genera constitute a family, and closely related families constitute an order.

Species-characteristic See Species-specific

Species recognition In ethological writing, an exchange of stimuli between the sexes that ensures that the partners are conspecifics, thus preventing the mating of individuals of different species. The concept thus applies to those elements of courtship behavior that govern the meeting of opposite-sex individuals of the same species, which accordingly serve to effect reproductive isolation. However, Mayr (1963) has rightly pointed out that the term "recognition" is misleading if it is taken to imply consciousness, as it does in ordinary talk about people; the lower animals have mechanisms for species recognition that are just as effective as those of higher vertebrates, yet most ethologists would dispute or dismiss the idea that such animals have consciousness.

An animal's recognition of species-identifying characteristics can either be inborn or result from early learning (for example, through sexual imprinting). Innate recognition is to be expected in those species that lack the opportunity for appropriate learning, as in most brood parasites and in the many insects and other invertebrates in which one generation dies before the members of the next hatch out.

Species-specific Characteristic of a particular animal species. Species-specific behavior is represented primarily by innate movement patterns (see FIXED ACTION PATTERN) and learning dispositions. Some, but not all, behavioral scientists reserve the term for features that, in addition to being characteristic of a species, are also exclusive to it. They thus distinguish "species-specific" from "species-characteristic" and "species-typical," which do not imply exclusiveness.

Species-specific behavioral characteristics have proven as useful for species identification as other characteristics and so can be considered taxonomic markers. However, difficulties are presented by species within which different populations possess different characteristics, as in the case of dialects, each of which is specific for only a part of the species.

Species-typical See Species-specific

Specific action potential(ity) (SAP) See Action potential

Spectral sensitivity   In the study of animals' visual capacities, the relationship between response threshold and light frequency. For example, the activity in an animal's optic nerve fiber can be recorded while shining light of different frequencies on its receptive field on the retina to find the minimum intensity of each frequency that will cause a change in the activity of the fiber. The result is usually a U-shaped curve, the low point of which indicates the part of the frequency spectrum to which the unit is most sensitive. The same approach can be taken to find the spectral sensitivity of a behavioral response by varying the colors of stimulus objects presented to elicit it, as has been done in studies of the pecking response of gull chicks (see STIMULUS SPECIFICITY).

Spectrum analyzer   See Sound spectrogram

Speculum   A white or colored patch on the body surface that contrasts with its background. German ethologists use the term for the white area around and below the base of the tail of many social mammals (ungulates, lagomorphs, primates), as well as for the often iridescent patches on the wings of birds, the denotation most common in English. The specula of ducks serve as a conspicuous signal for group cohesion, and males show them off by special movements during courtship. In the history of ethology they were among the first examples of intraspecific releasers.

The significance of the tail patches of mammals is still unclear and probably differs from species to species. In most cases it probably serves as a warning signal through special movements that signify the presence of a predator, as when a rabbit raises its tail in a "flashing" manner, which alerts the group and so triggers fleeing or hiding (see WARNING BEHAVIOR). Because this signal is conspicuous enough to draw the attention of the predator, and so increase the danger to the signaler, it has been conjectured that the significance of the tail flashing is not restricted to the warning context but also serves as a general appeasement signal within the group. However, this speculation is still in dispute.

Sperm   Semen; male gametes. Sperm are formed in the male gonads (testes), and as a rule, their discharge is accompanied by secretions from supplementary glands, for example, the prostate glands in vertebrates. The mass of sperm and secretions discharged at one time is sometimes referred to as the ejaculate. In some mammals part of the ejaculate congeals behind the sperm in the female's vagina, thus forming a sperm plug. In many invertebrates and some vertebrates (newts), the sperm are encased in a capsule called a spermatophore. In most animal species the sperm are actively moving cells propelled by undulations of thread-shaped tails (flagellae).

Spermatophore A capsule containing sperm. Spermatophores are either transferred by the male directly from his body to the body of the female in copulation or deposited on the ground or in some other place, and from there taken up by the female (see INDIRECT SPERMATOPHORE TRANSMISSION). In many species the spermatophores remain for a time in the female's body before the envelope splits, breaks up, or dissolves, releasing the sperm.

Sperm competition In species in which fertile females are subject to multiple matings with different males, the sperm of each male "competes" with that of the others for the goal of fertilizing the eggs, success usually depending on the order in which the matings occur. For example, in dung flies (*Scatophaga stercoraria*) the male who mates last fertilizes about 80 percent of the eggs, the remainder being more or less equally divided among the preceding males (see GUARDING BEHAVIOR). This contingency is referred to as sperm displacement. In contrast, in the spider mite (*Tetranychus urticae*) the first male to mate does all or most of the fertilizing. This pattern is referred to as sperm precedence.

Sperm plug See Sperm

Spontaneous behavior Autonomous behavior. Behavior that is internally caused and controlled and so neither released nor maintained by external or peripheral stimulation. A behavior pattern can be considered spontaneous if there is evidence "that there is no connection between the timing of its occurrence and the immediately preceding external stimulation" (Schleidt 1964). This definition does *not* imply that spontaneous behavior is completely free of stimulus influence; it says only that a specific command from the central nervous system to the musculature ensues that is independent of input from the exteroceptors. In practice it is difficult to judge whether an occurrence of a behavior pattern is truly "spotaneous," for the effective stimuli may escape the notice of a human observer, and even constant environmental conditions can elicit a response through cumulative effect after some interval of time.

Spontaneity has been attributed to locomotor movements and to the vocalization of various animals (turkeys and crickets, for example). Also the initiation of appetitive behavior is assumed to be spontaneous. Pronounced spontaneous movements are made by embryos, and such movements may have profound significance for normal behavioral development (see BEHAVIORAL EMBRYOLOGY). When a behavior pattern that is normally dependent on external stimulation occurs spontaneously as a consequence of extreme threshold change, the occurrence is described as vacuum activity.

Status signal   A recently introduced term for a sign that displays a specific socially relevant characteristic to conspecifics, such as position within a rank order or age or experience of a prospective mate. Thus the highest-ranking male in a group of gorillas displays a conspicuous silver-gray marking on its back (silverback male), in contrast to the black coloration of the rest of the members. In many songbirds there is evidence that color markings on the feathers signify rank position within a group (see BADGE OF STATUS). In other songbirds the older males that have more breeding experience possess larger song repertoires and in competition with other males acquire larger or better-quality territories. In the American red-winged blackbird (*Agelaius phoeniceus*), a polygynous species in which the size of the red wing patch of the male is a status signal, it has been shown that the males with better territories secure more females. An example of a temporary status signal is the often conspicuously fluctuating appearance in female primates indicating sexual readiness (see ESTRUS).

The main biological significance of status signals appears to be that they contribute to reduction of intraspecific aggression, for example, disputes over rank.

Stereotyped movement   See Stereotypy

Stereotypy Constancy of form in behavior; uniform repetition of motor patterns or vocalizations. Under natural conditions, fixed action patterns are highly invariant (see FORM CONSTANCY), as in the monotonously repeated cooing of doves and the rhythmic claw waving of fiddler crabs. Sometimes, however, learned behavior becomes considerably stereotyped, as is true of the songs of many bird species.

Ethologists apply the terms "stereotypy" and "stereotyped movement" to the rigidly fixed behavior patterns that are consequent on unnatural situations. They distinguish two categories of such compulsive movements. The first is the especially pronounced repetitve motions (such as jerking of the head) that animals and humans manifest as part of the deprivation syndrome. The other sort develops in captive animals as a consequence of confinement in impoverished conditions and consists of such odd stereotypies as head turning, swinging back and forth, somersaulting, walking or running back and forth along a fixed path, all repeated in rigidly uniform ways for hours on end. Mammals that normally live rich social lives are especially likely to develop such patterns when deprived of companionship and placed where they have nothing else to do. In zoo animals these patterns are sometimes associated with a single traumatic experience rather than with deprivation. Human children sometimes show stereotyped patterns, such as rocking movements with the head down and arms clasped about the knees, when confined to a hospital or as one of the symptoms of autism. In the stereotypy of animals conditioned by captivity, the orientation (see TAXIS) of the movement is "frozen," as it were, because the unvarying spatial relations require no adaptive variation. Scattered reports of compulsive movements like those seen in zoo animals have also been made of wild animals—elephants, polar bears, and chimpanzees. In primates stereotyped patterns, whether natural or due to captivity, may be handed on from one generation to another (see TRADITION). In situations of stress, lost stereotyped habits may be reinstated.

Steroid hormone A collective term for a group of hormones characterized by similar molecular structure. It includes the sex hormones of vertebrates.

Stimulation In ethology this term has several uses. In its broadest sense it refers to any effect of a stimulus on an organism or to the stimulus or stimuli themselves. If the stimulus emanates from a conspecific it is called social stimulation. In classical conditioning the unconditioned (or unconditional) stimulus serves as the reward (see REINFORCEMENT) for performances of the conditioned (or conditional) response.

Stimulus An external or internal physical condition or change that causes a physiological or behavioral response in an organism, for example the excitation of sense cells or nerve cells. In ethology the word is sometimes used as an abbreviation for sign stimulus, that is, for substantially more complex stimuli or stimulus combinations than is usual in other areas of biology. Compare IRRITABILITY.

Stimulus deprivation See Isolation experiment

Stimulus filtering Selection, from the total stimulation presented by a situation of those stimuli that elicit and regulate behavior (see RELEASING MECHANISM; SIGN STIMULUS). The filtering can occur in the sense organ, where the receptors exclude part of what they receive or block transmission of its effects (peripheral filtering). But as a rule at least some of the filtering takes place in the central nervous system, although in most cases little is known about where and how such central filtering occurs. It probably takes place in several successive steps (see SENSORY CODING).

Stimulus generalization Generalization. When an animal, after learning to react in some way to a particular object or situation, reacts the same way to objects or situations that are similar to but not identical with the conditions of the original learning, it is said to show stimulus generalization. The effects of the initial experience may be extended because the animal does not discriminate between the situations or because it perceives similarities and takes them to be what counts. If, however, the similarities are irrelevant, and the animal's behavior a consequence of its being made more reactive to any stimulation by the initial experience, the case is one of sensitization rather than stimulus generalization (see PSEUDOCONDITION-ING). Stimulus generalization occurs in many animal species to different degrees; within a species it probably varies in different behavioral contexts.

Stimulus generalization no doubt plays a role in many social recognition processes. An animal whose experience is restricted to only its mother or a few individuals (parents and siblings) during its first days or weeks of life later must be able to recognize a whole category of individuals, namely the members of its species or population, and distinguish them from members of other species or populations (see SPECIES RECOGNITION). Stimulus generalization probably also contributes to kin recognition. In some of the more highly evolved species the capacity for generalization attains a high level of sophistication in concept formation.

Stimulus modality   See Modality

Stimulus model   Surrogate; dummy used in experiments on the stimuli that elicit a specific behavior pattern. In this usage the word "model" has something of the sense it has when we talk, for example, of a scale model as opposed to a theoretical model or a mathematical model or an explanatory model. But a stimulus model need not be a lifelike replica of the object it refers to. Ethological experiments may use completely "unnatural" models to ascertain the characteristics of the essential sign stimuli or releasers governing some behavior pattern. Thus the model may imitate only parts of an animal (dummy beak, dummy head), or different features of the object modeled may be changed—enlarged, reduced, or presented in other shapes or colors. Even such unnatural objects as colored wooden balls or cubes are sometimes used. Reproduction of tape-recorded vocalizations or artificially generated sounds, as well as discharges of artificial scents can serve as stimulus models (see VOCAL SIMULATION). Experiments with stimulus models have played an important role in the research on animal behavior from its beginning.

There are also "natural" stimulus models. Many parasitic animals have, in the course of their phylogenetic history, imitated characteristics having a specific releaser function in another species so as to produce the same effect on the host (see BEHAVIORAL MIMICRY; MIMICRY). Thus the mouth markings of widowbird chicks (Viduinae) are exact copies of those of the young of the estrildine finch species in whose nests the brood parasite lays, and the larvae of many beetle species that live in ant nests secrete a pheromone ("false pheromone") so like that of the ant larvae that it releases the same parental care behavior.

Stimulus-response relationship The temporal and quantitative relationship between presentation of a specific stimulus pattern and the appearance of the appropriate behavior pattern. With simple reflexes the relationship is rigid and invariable. In more complex cases it may be graded, the kind or strength of the response varying with the intensity of the stimulus. Stimulus-response relationships may also vary with internal state, environmental context, and experience. For example, repetition of an inconsequential stimulus that initially evoked an orienting reflex may lead to its being ignored (see HABITUATION). Other matters of relevance here include threshold change, priming stimulus, conditioning, latency, and motivation.

Stimulus specificity Stimulus selectivity. The selective responsiveness to stimulation characteristic of a receptor or a behavior pattern. The "tuning" characteristics of physiological or behavioral response systems. For example, the retinal photoreceptor cells of vertebrates fall into several classes on the basis of differences in their spectral sensitivity. All receptors, with the possible exception of those for pain, can be specified, at least according to modality, as optical, chemical, and so forth. The kind of stimulation to which a receptor is most sensitive is referred to as the adequate stimulus.

Stimulus summation In ethology this usually refers to the additive effects of sign stimuli arriving together or in quick succession, as reflected in the strength or kind of response they elicit. For example, experiments using dummy eggs have shown, in the case of the egg-rolling response of herring gulls, that a larger egg is a stronger stimulus than one that is the same but smaller; that a speckled egg is more effective than an unspeckled one of the same size and ground color; and that a greenish egg is superior to a brownish egg. The most effective egg combines all of the positive characteristics. However, a deficiency in one characteristic can be compensated for by exaggeration of one or more of the others, for example a large, unspeckled egg can be as effective as a small, speckled one.

The collective effect and substitutability of sign stimuli were originally referred to as the phenomenon or law of heterogeneous summation, and this term is still used sometimes. However, quantitative studies have shown that the combined effects do not always conform to simple summations of separable independent contributions from the several stimuli concerned. For this reason it has been suggested that the less specific "mutual stimulus augmentation" (Curio 1969) may be preferable.

Besides its use in ethology, the term "summation" also occurs in neurophysiology, where an analagous pattern has been found in synaptic transmission, which is also described as "facilitation." Nerve impulses arriving in quick succession at one presynaptic terminal (temporal summation or facilitation) or more or less simultaneously at several presynaptic terminals (spatial summation or facilitation) can pool their depolarizing effects on the postsynaptic neuron sufficiently to fire it when one or two impulses would be insufficient to do so.

Stimulus threshold   See Threshold

Stochastic analysis   Statistical computation of the transition probabilities between items in mutually exclusive and collectively exhaustive sets, such as letters of the alphabet or the preening movements of a bird or fly. In applying such analysis to behavioral sequences (see SEQUENCE ANALYSIS), certain requirements have to be met such as that of stationarity, meaning that the transition probabilities must be stable, at least for the time covered by the sample of data (that is, the sequences must be ergodic). When the frequency of one item following another is greater or less than chance, the dependency is said to be Markovian or to conform to a Markoff chain model. If such a different-from-chance transition probability between two items is independent of the preceding items, the dependence is said to be a first-order Markoff chain; if the probability of B given A differs according to what immediately precedes A, the dependence is that of a second-order Markoff chain; and dependence extending farther back is labeled according to the same pattern.

Not all sequential patterns are stochastic in nature, that is, fully captured in terms of statistical transitions. Patterns governed by grammatical rules, such as those of language, which allow embedding for instance, escape detection by stochastic analysis. Another limitation is that stochastic analysis takes no account of variation in the duration of actions and the intervals between them, which can be at least as significant as the order of events in behavioral sequences. Even so, the approach has frequently been put to use in ethological study of behavior sequences, especially those of communication.

Strain   In genetics, a selectively inbred stock. Thus in work on *Drosophila,* mutant forms have been selectively bred to give rise to mutant strains, such as "ebony" and "vestigal winged," which are to a high degree genetically homogeneous.

Strategy   In behavioral science this term has acquired a specialized meaning. Formerly it was used only in roughly the same sense as "way of life." One still speaks of foraging strategies or reproductive strategies (monogamy, polygamy, family). But in sociobiology the term has become associated with mathematical game theory. Roughly speaking, "strategy" in this sense refers to a comprehensive behavioral plan to deal with decision making; for each decision situation it prescribes an action or at least specifies the probabilities of results of different actions for the different situations. The game theory of evolutionary biology poses the question of how far strategies for dealing with conflicts between two or more individuals can be understood as adaptive. It is also concerned with how natural selection "evaluates" strategies. Complicating an analysis of this sort is the fact that not only are strategies adapted to the world (nonliving and living) outside the species, they also concern situations between members of a species; in many cases the decision is affected by what others are doing. To bring calculation to bear on this problem, the concept of evolutionary stable strategy (ESS) has been developed. In the simplest case conceivable (assuming that most of the individuals of a population will "play"), one strategy is superior to all others. In this case an alternative strategy has no chance of becoming established or of displacing the first, which is therefore an ESS.

Stress   This term is used variously and is seldom precisely defined. In engineering it may refer to the physical forces that an object or an organism is subjected to, such as the g-force acting on a pilot pulling out of a dive. In ethology and population biology "stress" generally refers to the effect on organisms of environmental demands beyond the normal range. Such environmental pressures give rise to a characteristic pattern of physiological changes involving the nervous system and various hormones.

Recent investigation has shown that in higher vertebrates, what was earlier viewed as a unitary reaction is ascribable to two separate systems. A rapid adjustment (often on the order of seconds) to a distress situation calls forth an alarm reaction, involving activation of the adrenal medulla by the sympathetic nervous system. Secretion of hormones (adrenalin and noradrenalin) by the adrenal medulla causes a variety of changes, among which are acceleration of heartbeat, elevation of blood sugar level, increase of blood flow to the skeletal musculature, inhibition of alimentary tract activity, and heightened readiness for defense or escape behavior. This kind of adjustment is known as the flight or fight syndrome.

The other kind of change is caused by persisting environmental stress and is induced by increased secretion of hormones from the adrenal cortex, which leads to inhibition of hormone release by the thyroid gland and the gonads; as a consequence, growth can be retarded and sexual activity curtailed. These changes are referred to as the general adaptation syndrome (GAS), which can eventually lead to disease states such as stomach ulcers. The injurious factors in such situations are called stressors. These may be environmental in nature (heat, cold, lack of food or water, oxygen deficiency, infection, poisoning) or may have their source in conspecific stimulation, such as occurs in fighting and other consequences of high population density (social stress). It has also been found that within a species—no doubt as a consequence of genetic variation or previous experience—some individuals react to stress situations mainly with the flight-or-fight syndrome (mobilization), others react more with the general adaptation syndrome (depression), and still others display both syndromes more or less equally.

The general adaptation syndrome, as the name implies, was originally thought to involve only "general" or unspecific reaction. That notion is now disputed. Various findings indicate that a certain degree of specificity obtains; in particular, the adrenal cortex can vary its hormone output as a function of differences of input (either from the sympathetic nervous system or from hormones from the pituitary). Because of its significance for the human case, considerable attention has been given to the results of social stress, especially with regard to reproduction. In addition to physiological effects (reduced fertility in males, reduced fecundity in females, increased infant mortality, delayed sexual maturation) social stress can cause behavioral changes (reduction or cessation of parental care, cannibalism, infanticide, reduced curiosity and learning, and much else).

Stress symptoms are all too common in captive animals where housing conditions are inadequate (for example, too many individuals in too small a space). Although under natural conditions pronounced stress phenomena are seldom seen, the early stages of social stress are not uncommon and may serve as an important means of regulating population density, counteracting increases that would lead to numbers too great for the available resources to support.

Stressor   See Stress

Stridulation   See Sound production

Submissive behavior   Submission; appeasement behavior. Performance of an appeasement gesture by the defeated animal in an intraspecific conflict. Such behavior can have such a strongly inhibiting effect on the victor that the struggle is terminated. The form of the behavior is commonly the reverse of that in a threat display (see THREAT BEHAVIOR), a relationship that Darwin labeled the "principle of antithesis." Thus instead of standing as tall as possible, the defeated animal may lower itself to the ground; instead of facing toward its opponent it faces away and thus "turns off" the threat signal.

Submissive behavior is seen especially in pugnacious animals such as dogs and wolves. By acting submissive, the animal can avoid serious injury, even when there is considerable difference in the strength of the contenders.

Subsong   A form of bird vocalization that seems to be functionless, for it appears to have none of the effects usually associated with song, such as territorial advertisement or mate attraction. But it is improbable that it is completely without function. Subsong differs from "functional song" or "motif song" in being, as a rule, of lower intensity, more variable, and wider in frequency range, and in having more and longer components and a virtual absence of consistent motifs. It is most common in early spring before the onset of the loud motif song, and in autumn during a short period of resurgent gonad activity. In many bird species, however, motif song and subsong may occur side by side. The young males of many species go through a stage prior to the first utterances of motif song when they sing a subsong that is referred to as juvenile song.

Quite differing views have been expressed on the significance of subsong. Possibly it serves as exercise or practice, at least in young birds. If so, it conforms in this and in some other respects—great variability, free recombination of elements, occurrence at times when no other behavioral tendencies are activated—with the characteristics of play behavior. It has therefore been repeatedly classed with play behavior and also described as play song.

Substitute activity   See Displacement activity

Substitute object   A "makeshift" object to which a behavior pattern is directed in the absence of its usual object, as a consequence of threshold change. Sometimes the object of a redirected movement (see REDIRECTION) is also referred to as a substitute object, but because this situation is distinctly different from the first, the term "neutral object" is perhaps preferable for this case.

Subthreshold stimulus   See Threshold

Summation   See Facilitation. Addition of the effects of two or more neural transmissions. There are two forms of summation: temporal and spatial. In temporal summation impulses arriving successively at the same axon terminal progressively increase the probability of firing (or inhibiting) the postsynaptic nerve cell; that is, they progressively depolarize the cell membrane until the threshold is reached for generation of an action potential. In spatial summation excitations arriving simultaneously via different axon terminals, which would be ineffective singly, together produce sufficient change to cause response, for example across a synapse or at an effector organ.

The concept of summation has also been applied to the elicitation of behavior patterns, as in the "law of heterogeneous summation" (see STIMULUS SUMMATION).

Sun-compass orientation   See Orientation

Supernormal stimulus   A stimulus that releases a specific behavior pattern more strongly than the natural "appropriate" sign stimulus or releaser or that is preferred to the latter in a situation of choice. Supernormal stimuli were first discovered in experiments using artificial stimuli (see STIMULUS MODEL) that exaggerated certain features functioning as releasers, with the result that such stimuli worked better than true-to-life imitations of the features. However, supernormal stimuli have also been found in nature, the most familiar example being the gaping of young cuckoos, which is more conspicuous and thereby surpasses that of the host nestlings in getting the attention of the parents (see INTERSPECIFIC RELEASER). Also, some "guests" harbored by ants in their nests have characteristics that stimulate their hosts more strongly than the ants' own corresponding social signals (see BEHAVIORAL MIMICRY).

In earlier ethological literature the term "superoptimal stimulus" was sometimes used in this sense. However, it is literally nonsense, because "optimal" already means the best possible; it is now hardly ever used.

Surplus hypothesis   See Displacement activity

Surrogate   A term used mainly in the experimental literature on primates, where it is synonymous with dummy or stimulus model. The usual contrivance to which the term is applied is an object substituted for the mother (so-called mother surrogate) in experiments involving the hand rearing of young (see ISOLATION EXPERIMENT).

Survival value   Adaptive significance, biological function. Survival value is the answer to the question "What is it for?" asked of a species characteristic. It is the way that characteristic contributes to survival and reproductive success. Survival value specifies the selection pressure or pressures to which the characteristic answers. For example, experiments on black-headed gulls confirmed that their practice of carrying away eggshells from the nest shortly after hatching has the survival value of defending the clutch and brood against predation: the conspicuousness of broken shells might attract the attention of crows and other animals that prey on the eggs and newly hatched chicks; removal of the shells avoids this risk.

Symbiosis An enduring arrangement of two organisms of different species living together. There is some inconsistency in the use of the term. Some writers (such as Wilson 1975) include all interspecies cohabitation irrespective of how the two organisms are affected, and distinguish three kinds of symbiosis: mutualism, in which both partners benefit from the relationship; commensalism, in which one partner benefits but the other derives neither benefit not cost; and parasitism, in which one partner exploits the other (the host) to its detriment. Other writers consider symbiosis synonymous with mutualism. Symbiosis in this more restricted sense occurs between plant and plant (fungus and alga in lichen), between plant and animal (flower pollination by insects), and between animal and animal (anemone fish and sea anemone, cleaner fishes and their "clients"). A very simple form of symbiosis is probably presented by mixed flocks, such as those of many species of birds (see MIXED-SPECIES GROUP).

Mutualism involves numerous adaptations of body form and behavior. Thus when a hermit crab is forced to leave a shell it has outgrown, it detaches the symbiotic sea anemones from the discarded shell and carries them in its chelipeds to the new shell it has selected to occupy. A form of symbiosis in which behavior plays an especially important role is cleaning symbiosis.

Symbolic movement Symbolic action. A term used mainly in the earlier ethological literature, but without consistency. Lorenz (1941) applied it to intention movements that have undergone ritualization and so serve as signals in communication between conspecifics. Other authors used the term in a wider sense that was roughly the same as "display." An example of what was referred to as symbolic movement is the tidbitting that is a component of courtship in many gallinaceous birds. Nowadays the term is seldom used.

Sympatric Being in contact geographically; the opposite of allopatric. Populations and species are sympatric when their ranges of distribution overlap and they have at least the opportunity for interbreeding. If interbreeding occurs, either genetic heterogeneity between the populations will be eliminated through gene flow, or some means of effecting reproductive isolation between them will be selected as a consequence of the poorer reproductive success of hybrid matings. Such selection was once believed sufficient to bring about sympatric speciation. However, evidence now indicates that some period of allopatric separation is necessary for genetic differences to accumulate to the point where interbreeding will be selected against intensely enough to counteract the homogenizing effects of gene flow.

Pair formation behavior has frequently been affected by selection for reproductive isolation by exaggeration of differences and sharpening of discrimination between the populations.

Synapse Communicating junction between two nerve cells, between a receptor cell and a nerve cell, or between a nerve cell and a cell of an effector organ (muscle or gland). In most cases the transmission can go only one way, from the presynaptic cell to the postsynaptic cell; the membranes of the two cells are separated by a microscopic space (the synaptic cleft) into which the presynaptic membrane discharges a quantity of transmitter substance on arrival of a nerve impulse. Uptake of the transmitter substance by the postsynaptic cell may locally reduce the electrical difference between the inside and outside of the membrane (depolarization), thus lowering the threshold to further such transmission (threshold change); or, in an already primed cell, depolarization crosses the threshold for firing and thus generates a nerve impulse in the postsynaptic cell; or it locally increases the electrical difference at the membrane (hyperpolarization), thus elevating the threshold to further transmission. The latter case is an inhibitory synapse.

Synapses may be categorized according to the transmitter substance with which they are associated, for example, adrenergic, seritonergic, or cholinergic according as the transmitter substance is adrenalin, seritonin, or acetylcholine. However, synaptic transmission is not always chemical in nature; sometimes the arrival of presynaptic impulses causes direct electrical change. Such cases are described as electrotonic synapses. They are proving to be more common than was once thought.

Synchronization   Temporal coordination. As far as behavior is concerned synchronization is of special significance in two main connections: with periodic aspects of the nonliving world and between conspecifics (and, in exceptional cases, between members of different species). In the first case the environmental patterns are, usually, daily and yearly cycles, which affect behavior in numerous ways (see ANNUAL PERIODICITY; CIRCADIAN RHYTHM). The synchronization depends upon a zeitgeber, among other things. In intraspecific processes, temporal coordination is achieved through various forms of communication. Such synchronization plays an especially important role in preliminaries to copulation and in the behavior of closed groups. Thus an essential function of courtship is to effect the physiological and behavioral tuning of each partner to the other. Within a group social facilitation can contribute in decisive ways to synchronization. Remarkable cases of timing that combine environmental and intraspecific factors include the annual breeding eruptions of palolo worms and of grunion fish; in both species the animals perform precisely on cue with a particular phase of the moon.

Syndrome   Symptom complex. In ethology the term is used mainly in reference to the deprivation syndrome, which encompasses all the developmental abnormalities consequent on experience deprivation (see ISOLATION EXPERIMENT). In addition writers sometimes refer to a behavior syndrome, meaning a larger behavioral category made up of several categories of smaller scope.

Synecology   See Ecology

Synergism   Working together; combined action, as when two hormones have an effect jointly that neither would have if present alone (for example, estrogen and progesterone in the formation of brood patches in many birds). The word contrasts with, but is not exactly the opposite of, antagonism.

Syntactics   See Communication theory

Syntax In linguistics, the rules governing how words are combined to form phrases, clauses, and sentences; it is thus part of grammar. By analogy the term has been applied to the combining and ordering of signals in animal communication, especially where the combinations may convey different information or produce different effects from those of the signals used in isolation.

TERRITORIAL SONG

T

Taming   A process that progressively reduces and eliminates a wild animal's initial tendencies to flee and to react negatively toward humans. It leads to the condition of tameness, which is achieved when the fleeing tendency has disappeared; sometimes a stable positive relationship to one or more humans is established. Taming is thus to some extent comparable to socialization, although it involves orientation toward a biologically "wrong" object. In the literature taming is sometimes confused with imprinting in that the two terms are used interchangeably or at least are not clearly differentiated. Imprintinglike fixations can certainly contribute to taming, especially when taming begins at a very early age, but there is nevertheless an essential difference. With imprinting to humans (see ERRONEOUS IMPRINTING) the animal directs sexual or aggressive behavior toward the object on which it is fixated. Taming, on the other hand, involves merely the absence of negative reaction (fleeing), and possibly the occurrence of neutral positive reactions such as staying in the vicinity of a person or approaching for feeding. The processes of habituation and conditioning contribute to the establishment of tameness. Compare DOMESTICATION.

Target organ A part of the body upon which a hormone acts to produce temporary or lasting change in ontogenetic development, seasonal fluctuation, or sensory, neural, or glandular activity. For example, the gonadotropic hormones secreted by the pituitary have as their target organs the gonads, whose maturation, activation, and secretory activities they help regulate.

Taste aversion Bait shyness. A form of avoidance conditioning. As a consequence of experiencing noxious or poisonous effects from a particular kind of food, an animal can learn to avoid ingesting it again. This kind of associative learning is unusual in that the consequences of the action may not be experienced for an hour or more after its occurrence; in other kinds of learning the shorter the latency between response and reinforcement, the stronger the effect. Many Batesian and Mullerian mimics (see MIMICRY) and their models depend upon taste aversion for the protection from predation that their appearance provides.

Taxis Orientation reaction. Reaction oriented to external stimulation; movement having to do with spatial orientation. The orientation of some animal movements was once likened to the way plant growth is influenced by the direction of light and the pull of gravity, a phenomenon known as tropism; for a time that term was applied to animal oreintation. However, differences and variants in animals became apparent, which led to the decision to reserve "tropism" for stimulus-oriented growth (of which there are instances in animals, such as the growth patterns of hydroid colonies and corals) and to use "taxis" for oriented animal movements. Originally the term was applied to oriented movements in general, but because most of an animals' movements are oriented in some way, this broad application has limited usefulness. The definition now is usually narrowed to oriented turning within a movement sequence. However, many authors also include the orientation of a stationary animal when the angle of the body axis with respect to a stimulus source is an "actively held" stance (Schöne 1973), even though the posture may appear passive.

Spatial orientation is aided by different sense modalities, as indicated by the corresponding terms. For example, orientation to light is called phototaxis; that to chemical stimulation is called chemotaxis; and that to gravity is geotaxis. If the orientation is toward the stimulus source, it is called a positive taxis, and away from the source a negative taxis. Thus if an adjustment in direction is toward a light source, the animal is described as positively phototactic. When the direction of movement is at some other angle to the direction of the stimulus source, the orientation is referred to as menotaxis. The best-known example of this is sun-compass orientation. Orientation based on memory of the path to a place is sometimes referred to as mnemotaxis.

The mechanisms underlying spatial orientation also present a basis for differentiation. The most familiar to us is orientation guided by perception of a distant object, known as telotaxis. This entails patterned perception of the sort represented by advanced visual systems; other examples are the sonar of bats, the infrared "vision" of certain snakes, and the electrical sensitivity of certain fishes. A less precise means of determining direction uses comparison, by two symmetrically placed sense organs, of the intensities of stimulation on each side. When these intensities are unequal, this process, called tropotaxis, induces the animal to make a compensatory turning movement. A third form, known as klinotaxis, uses a single sense organ that moves about to provide *successive* comparisons of stimulus intensities from different directions. Thus fly larvae find dark places in which to pupate by swinging their light-sensitive anterior ends from side to side as they creep along, turning toward the darker side when there is a difference.

All of the orientation reactions mentioned so far are directed in relation to the stimulus source and are collectively known as topotaxes. In contrast are undirected reactions to change of intensity of stimulation. German ethologists sometimes refer to these as phobotaxes, even though, being unoriented, they are not true taxes. The more frequently used term is kinesis. Klinokinesis, for example, works by the principle of trial and error, the animal changing direction whenever it finds itself in an unfavorable situation (in relation to humidity, say, or salt concentration in water). By this means it can become hedged in a zone of most favorable environmental conditions. Such behavior is shown by paramecia and wood lice.

Taxis component   A term occurring mainly in the earlier ethological literature, where it refers to the stimulus-dependent oriented component (see TAXIS) of a behavior pattern. Within a complex behavior pattern one can often differentiate between a relatively inflexible fixed action pattern, whose form is largely independent of external or peripheral stimulation, and flexible movement, whose course is dependent on such stimulation. The latter elements, which serve to adjust the orientation of action with respect to its object, as in prey catching, are described as taxis components. They may precede the fixed action pattern (orienting the body axis in the direction of the prey before moving to seize it) or may simultaneously combine with it in various ways. A classic example of the latter is the egg rolling of such ground-nesting birds as the greylag goose.

Teleology See Teleonomy

Teleonomy The scientific investigation of programmed purposiveness (Osche 1973) in living organisms and in certain manmade machines. Behavior patterns are quite prominently teleonomic; that is, they often appear "goal-directed" yet can be fully accounted for in terms of antecedent proximate causation. Teleonomic behavior is governed either by a genetically determined (closed) or experientially acquired (open) program, and through this program behavior is adjusted in fine detail to the prevailing environmental conditions to arrive at some end state; in this sense the behavior is purposive. Such adjustment is often achieved through the guidance furnished by feedback. Study of evolutionary adaptation through selection has helped us understand how functional design can result from purposeless process.

Nevertheless, the manifest functionality of animal behavior has frequently led to obscurantist teleological interpretation, which invokes goals as causes. The goal-directedness of the behavior is taken to imply consciousness and insight, attributes whose existence is a matter of conjecture (see ANTHROPOMORPHISM). In many cases such interpretation is unwarranted, for the teleonomic characteristics of behavior can originate through natural selection in the same way as any other adaptive characteristic. However, this natural selection is never directed toward a "goal" but represents an *a posteriori* process, genetic mutation providing the basis for improvement, which subsequently "rewards" the individuals so endowed by enhancing their reproductive success (see FITNESS). The teleonomic explanation of behavior contrasts with a teleological interpretation in being essential causal and mechanistic. It says merely that each kind of behavior has a biologically significant "goal" in relation to the environmental conditions and that this relationship has arisen in the usual way through natural selection.

Telotaxis   See Orientation

Template   A term used in bioacoustics for information placed in memory that provides a model against which an animal can match its own production during vocal development. In many songbirds the young males have to learn their songs, often quite early in life, from the father or other exemplars. This learning often occurs far in advance of initial song production, in many cases during a limited sensitive period (see SONG IMPRINTING). At this time the young animal acquires a template, or "sound print," of the exemplar by ear, to which it compares its auditory feedback in shaping its own song during later development. Artificially deafened songbirds, which cannot hear their own song and therefore cannot compare it to the template, develop very atypical songs as a rule. For bird species in which the song is at least partly genetically programmed, some authors talk of an innate template that may be completed or modified through experience.

Tendency   See Action potential

Territorial behavior   Behavior patterns used for advertisement and defense of territory. They include marking behavior, threat, and territorial fighting.

Territorial flight   See Courtship flight

Territorial marking   See Marking behavior

Territorial song   Singing that advertises possession of a territory to rivals of the same sex, a form of marking behavior. Territorial marking is probably the main function of song in most songbirds and many other animal groups. But much territorial singing also serves other functions as well, such as attracting unmated partners prior to pair formation and stimulating and synchronizing reproductive activity afterward.

Territory   The broadest and perhaps still the best definition of "territory" in animal behavior is "any defended area" (Noble 1939). Because territories and territorial behavior take so many forms, it is difficult to be more specific without omitting some. Most commonly a territory is an area of more or less fixed boundaries from which the individual or individuals in possession exclude all rival conspecifics, or at least attempt to do so, by means of territorial advertisement (vocalizations, chemical signals), threat, and, if necessary, territorial fighting. However, some territoriality is centered on a moving location, as in the case of the bitterling, which lays its eggs within the mantles of a free-living mussel; defending its territory necessitates moving wherever the mussel goes. Territorial exclusion may apply only to conspecifics of the same sex (with the possible exception of same-sex offspring), but in some cases the exclusion includes all conspecifics and may even extend to members of other species (see INTERSPECIFIC TERRITORIALITY). If the territory is the possession of a pair or a group of animals, the members of the group show tolerance toward one another. Territoriality has been described for all classes of vertebrates and also for some invertebrates (crabs, spiders, insects).

Territories vary in size and function. Some are large enough to be the source areas of food for the owner and its offspring (feeding territories) and may or may not also provide the breeding site. Some are less than a meter square and may serve only as a place for display (as in lek species), courtship and mating (pairing territories), or nesting (nesting territories). Other functions include avoiding disturbance by conspecifics, which would interrupt the course of reproductive activities, and acquiring knowledge of the local geography to facilitate finding food and places to flee to in the face of danger.

Acquisition of territory involves intraspecific aggression because the boundaries are established chiefly through defense, but these boundaries are later respected without much dispute, at least between established neighbors. Merging with the concept of territory in one direction is that of individual distance, and in another, that of home range. Individual distance is distinguished from territory in that it has no reference to anything, stationary or moving, apart from

310

the animal's own body; and home range differs from territory in implying nothing about defense. However, the distinction between territory and home range, and between active defense and passive avoidance, is not always clearly drawn; and a single area may be defended in differing degrees during the year. The original territory concept is tending to fall more and more into obsolescence, its place taken by a description of the degree of exclusiveness with which the resources of a specific area are utilized by its inhabitants.

Testosterone   See Androgen

Thanatosis   Playing dead; death feigning; tonic immobility in the presence of a predator. Some predators will treat an animal as prey only if it is moving. By becoming motionless, an animal attacked by a predator can thus try to avoid being killed. It is usually a last-ditch resort, after attempts at concealment, fleeing, and defensive fighting have failed (see CAMOUFLAGE; DEFENSIVE BEHAVIOR). At least in some cases thanatosis appears to be induced by hypnosis. Death feigning has been observed in a number of insects, spiders, birds, and mammals.

Thinking   Mental activity of a logical nature. Many people consider the concept inapplicable to animals other than humans (see BEHAVIORISM). However, observational and experimental evidence, such as results from tests of problem-solving ability, indicate that at least the higher vertebrates, and even some invertebrates (honeybees) are capable of inferential reasoning. See COGNITION; INTELLIGENCE.

Threat behavior   Behavior patterns for intimidating or repelling an opponent in a situation likely to lead to combat. Common forms of threat include bristling of hair, ruffling of feathers, and raising of folds of skin or crests, all of which enlarge the body contours; and prominent display of "weapons" (teeth, horns), often combined with intention movements of the actions used in actual fighting. At the same time elements of fleeing behavior may be shown, indicating that fleeing as well as attack tendencies play a role in threat. The same patterns of behavior may also occur in courtship, where the context qualifies their meaning so that they play a role in pair formation. In many insects and vertebrates, sounds are used as threat signals. Threat behavior has a spacing-out function similar to that of actual fighting, yet with less expenditure of energy and less risk of physical damage to the contestants. It can occur in interspecific encounters, as when prey capable of defending themselves face a predator.

Threat face   A kind of threat behavior deriving from the development of facial musculature (see FACIAL EXPRESSION). A familiar example is the baring of canine teeth, which has evolved independently at least three times in mammals: in primates, carnivores and in cervids. Primates and carnivores draw back the corners of the mouth and thus reveal both the upper and lower canines. In many species the effect is enhanced by the extreme opening of the jaws (threat yawn or threat gape). Cervids draw back the upper lip and so bare the upper canines, which project like tusks in the more primitive species (musk deer, muntjacs). The same threat movement is made by some of the more advanced cervids also, such as red deer, in which the elaboration of antlers has been accompanied by a regression in the development of the canines to the point where they are no longer of use in fighting. These animals thus threaten by flourishing "weapons" they no longer possess, and the movement makes sense only in terms of its evolutionary origin (see BEHAVIORAL PHYLOGENY; RELICT).

Threat yawn   See Threat face

Threshold   The minimum strength of a stimulus necessary to elicit a response, such as a behavior pattern or a nerve impulse. Also the lowest perceivable level of a stimulus; for example, the auditory threshold of a specific tone is the lowest amplitude level of that tone that is just discernible. Stimuli below this value, which produce no perceptible effect, are referred to as subthreshold stimuli. In some cases subthreshold stimuli can act as priming stimuli. Threshold levels are usually not fixed in value, but vary with internal and external conditions.

Threshold change   Increase (threshold elevation) or decrease (threshold lowering) of the intensity or specificity of stimulation necessary to elicit a neural or behavioral response. In the case of a neuron, for example, immediately after firing the threshold is raised so high that even the most intense stimulation is insufficient to fire it again, and the cell is said to be in a refractory period. One or more subthreshold stimuli can lower the threshold of a resting neuron to the point where the next synaptic discharge will be sufficient to fire it (see PRIMING STIMULUS; STIMULUS SUMMATION).

Analogous changes occur with regard to behavior. Immediately after performance of an end act, it is difficult if not impossible to release the act again or elicit the associated appetitive behavior, but as time passes the threshold tends to steadily decline, so that less and less specific stimuli are sufficient to release response; in the extreme situation the threshold apparently reaches zero and the behavior occurs spontaneously (see VACUUM ACTIVITY). A familiar example is the head-shaking movement that domestic dogs perform with slippers, bits of cloth, or other "substitute prey" (see SHAKING TO DEATH).

Threshold lowering See Threshold change

Thwarting In ethology this term usually refers to a situation in which appetitive behavior fails to achieve the stimulus conditions necessary for completion of the activated behavior sequence. More generally, any situation where ongoing behavior is frustrated in reaching its proximate goal by absence of the requisite stimulation. Its effects are similar to those of conflict among incompatible tendencies in that it can be the occasion for displacement activity, for example.

Tidbitting A display by gallinaceous birds, originally consisting of pecking at the ground and calling, used for enticing chicks to approach and feed. In many species the behavior has been further ritualized and incorporated into courtship patterns, where it serves the secondary function of luring the female to the male.

Analogous behavior is found in many species of birds having nidifugous young and is often referred to as food calling or food luring. Here too what was originally a means of getting chicks to find food for themselves has frequently evolved into a courtship pattern whereby the male entices the female to approach.

Time giver See Zeitgeber

Time sharing As used by McFarland (1974) in connection with alternation between activities, a situation in which one of the activities controls when the other can occur. When the causal system for the controlling or dominant activity is in operation, it inhibits that of the other, subdominant activity, which can be expressed only during pauses in the dominant activity, when the subdominant activity is disinhibited. Interruption of activities can indicate which one is dominant in a time-sharing situation. If the interruption occurs during performance of the subdominant activity and does not extend past the point of recurrence of the dominant activity, the temporal organization of the behavior will not be affected—the dominant behavior will resume on schedule. If the interruption occurs during the dominant activity, the next pause in that activity will be correspondingly delayed, thus altering the timing of the subdominant activity. Experiments using this approach are called masking experiments; they have produced evidence of time sharing in a variety of contexts, such as courtship in sticklebacks and feeding and drinking in doves.

Tissue A structural and functional fabric made up of cells of more or less the same histological type bound together in a mass or sheet. Complexes of tissues constitute organs. Examples include muscle tissue, skeletal tissue, epithelial (lining layer) tissue, and glandular tissue.

Tool use The manipulation of objects for purposes such as care of the body or acquisition and preparation of food. Tool use has been described for a large number of animal species and in connection with many different kinds of behavior. Hermit crabs slip into empty gastropod shells, which they carry around as a means of protection for their vulnerable abdominal segments; the Egyptian vulture throws stones at ostrich eggs to smash them and get at their edible contents; elephants hold sticks with their trunks to scratch their backs; and chimpanzees have been seen to use sticks to strike at leopards or pile boxes on top of one another to reach food hanging out of reach. The best-known cases of tool use include that of the Galapagos woodpecker finch, which takes a cactus spine in its bill and pokes it into bark to get insects, and the North American sea otter, which breaks open shellfish by holding them in its forepaws and dashing them against a stone cradled on its chest as it swims on its back. Still further examples have been reported for wasps, blue jays, and baboons.

Sometimes even conspecifics are used as tools; weaver ants use their mandibles to pick up larvae, whose silk glands they apply to the edges of rolled-up tobacco leaves, working the larvae back and forth like weaving shuttles to "sew" the edges together. As adults the ants no longer possess silk glands.

Some animals go so far as to fashion or improve the tools they use. Thus chimpanzees break off twigs from which they peel the bark and remove side branches before poking them down holes in termite nests to fish for termites. Also the woodpecker finch can break off the cactus spine it uses as a probe.

Cases have been reported of animals using objects as tools in novel ways apparently arrived at by perception of configural relationships rather than trial and error. Such cases have been cited as instances of insight learning. There are also cases of a particular kind of tool use being handed on from one generation to the next through tradition.

Topotaxis   See Taxis

TOTE   Acronym from Test-Operate-Test-Exit, which refers to a unit of control in a hierarchy conception of how behavior is functionally organized. The term was introduced in the book *Plans and the Structure of Behavior* by Miller, Galanter, and Pribram (1960). The unit consists of a feedback cycle in which a mismatch (Test) between what an action is aimed at and what it achieves leads to repetition of the action (Operate), another comparison between aim and result (Test), which, if it shows a match, leads to moving on (Exit) to the next operation in the sequence. The hierarchical grouping of such units organizes action efficiently in the realization of "plans." This conception builds on the reafference principle in much the same way as Tinbergen's (1951) hierarchy theory built on Lorenz's (1950) conception of the mechanism governing instinctive activity.

315

Trace fossil   See Fossils

Tradition   Cultural inheritance. Transmission of acquired (learned) patterns of behavior or information from one generation to another within a group of animals. It is a prerequisite for any culture. In behavior studies, distinctions have been drawn within two different viewpoints to give four forms of tradition: direct and indirect, object-bound and non-object-bound. With object-bound traditions the experienced and inexperienced individuals must come together with the object (food, say) about which knowledge is to be acquired or its treatment learned. In the non-object-bound form the information about the specific object—perhaps the location and quality of a food source—is transmitted even in the absence of the object by some form of symbolic representation, as in the dance language of honeybees.

The second basis of differentiation has to do with the relationship between the partners in the transmission of tradition. In the direct form the neophyte learns, for example by imitation, in the presence of the instructor, usually an adult. In the indirect form the inexperienced animal is placed in a situation where the instruction can occur without the experienced animal. For example, many insects transmit oviposition preferences to their offspring by laying their eggs on the plants of choice; when the young mature they lay their eggs on the same kind of plant as that on which they fed as larvae. Thus the preference for a plant species is handed on from generation to generation without the members of succeeding generations ever meeting. The only prerequisite for this form of tradition is brood provisioning. Indirect traditions can only be object-bound, obviously; direct traditions may be either object-bound or non-object-bound.

In contrast to the genetic transmission of information, tradition allows for the transmission of information to any number of individuals, at least within a group. It can thus lead to rapid diffusion of a behavior pattern. A well-known example of this is the potato washing in an island population of Japanese monkeys (red-faced macaques). The behavior was "invented" by a young female, then spread gradually over the entire island as a behavioral modification. Tradition also plays an important part in the transmission of dialects and in the acquisition of travel directions, such as the migration routes of migratory birds. In cases where individuals of one species take over patterns, such as the vocalizations, of another ("mocking") it is possible to speak of interspecies tradition.

Transmission of information by tradition is especially likely to occur where—as in almost all nonmonogamous primates—three generations are represented in a group.

Training A broad term for learning procedures (very different from one another in details) controlled by humans. In the course of training, animals acquire some humanly designed behavior pattern, have an inherent behavior pattern built up through coaching, or achieve some goal set by the trainer (see SHAPING). Training techniques are also used to get rid of undesirable behavior by punishing the animal for its occurrence or rewarding for its withholding. As with conditioning, which is synonymous with training in many contexts, training usually employs reinforcement. In training, circus animals or dogs, the direct relationship between trainer and animal plays an important role. In training an animal for scientific investigation, on the other hand, such a relationship is usually avoided as far as possible. In behavioral study, training techniques provide means of investigating sensory capacities, orientation mechanisms, and learning abilities.

Transitional action Postural facilitation. When a movement occurs in the same or similar form in two different functional systems (see MULTIPURPOSE MOVEMENT), its performance in the service of one system can lead to a switch in motivation to that of the other. Also, in situations of conflict or thwarting, the behavior performed may put an animal into an attitude similar to one that leads into some other kind of behavior, which is thus promoted and occurs as displacement activity. For example, the threat display of a male three-spined stickleback places it in a head-down position similar to that used in nest digging; this, together with the stimulation of the close proximity of the head to the sand, may be the reason why displacement sand digging occurs in this situation. An alternative conjecture is that the displacement digging came first, the threat display being a ritualized derivation of it (see RITUALIZATION).

Transition probability An estimation of the likelihood that given behavior pattern p, behavior pattern q will follow. Such estimations can be made from tallies of the occurrences of sequence combinations in samples of observations of behavior. The actual transition probabilities can be compared with those predicted by chance to find the extent to which there might be stochastic dependence in the sequences (see STOCHASTIC ANALYSIS).

Transmitter substance A chemical discharged by the arrival of a nerve impulse at a synaptic terminal, effecting synaptic transmission. The substance is taken up by the postsynaptic membrane, which it consequently either depolarizes, promoting firing of the postsynaptic cell, or hyperpolarizes, inhibiting the postsynaptic cell. The most common transmitter substances are acetylcholine and epinephrine, but a variety of others, specific to certain sites or systems, have recently been discovered in vertebrate brains. Synapses at which acetylcholine is the transmitter substance are called cholinergic; synapses at which epinephrine is the transmitter substance are called adrenergic (after adrenalin, an alternative name for epinephrine); the terms for synapses involving the other transmitter substances follow the same convention.

Transport of young Carrying of young by one or the other or both parents, either inside or outside the parent's body. Examples of animals that transport within body cavities (mouth, pouches, skin sacs) are the mouth-breeding fish, many frogs and toads, and marsupials. External transport occurs in many spiders, scorpions, some swimming birds (swans, great crested grebes), koalas, sloths, anteaters, and all primates. The young of primates, in which the behavior is especially pronounced and associated with close bodily contact with the mother (the father in marmosets), are described as clinging young.

Transient transport of the young by the mother is exemplified by the retrieving of rodents and feline carnivores (see CARRYING IN).

Treptics See Apotreptic behavior

Trial-and-error learning See Conditioning

Triumph ceremony A behavior pattern observed in greylag geese, and comparable patterns in other geese. The male makes a sham attack on a rival male, then returns to his female, giving the "triumph" call and threat displays toward a point beside her. The female may call in unison with the male, in which case they have very likely established a pair bond. The display sequence thus serves to establish and maintain the bond. The detailed investigation of this ceremony in greylag geese has led to its interpretation as pair-bonding behavior expressive of a bonding drive that is independent of sexual motivation.

Trophallaxis The transfer of gut contents from one individual to another. It is common in social insects. Such food dispensing from either the mouth or the anus, can be mutual or unilateral. It can occur between colony members or between a colony member and a member of another insect species kept and exploited as "farm animals." Among colony members it can serve as a means of chemical communication (see PHEROMONE).

Tropism Regulation of growth direction in organisms attached to a substrate, that is, plants and sessile animals. Earlier the term also referred to the orientation of locomotor movement relative to a stimulus source, but this sense is now consigned to the term "taxis." Tropisms are differentiated according to the nature of the environmental agency to which they correspond. Thus growth governed by light is phototropism; that governed by gravity is geotropism. Often the growth pattern is subject to the influence of more than one factor, the commonest combination being light and gravity.

Tropotaxis See Orientation

Typical intensity In the processes of ritualization and emancipation, a display is likely to change from a graded response (one that varies with intensity of the stimulation eliciting it) to a more or less fixed pattern for most of the range of stimulus intensity to which it answers. Such constancy of form was dubbed "typical intensity" by D. Morris (1957).

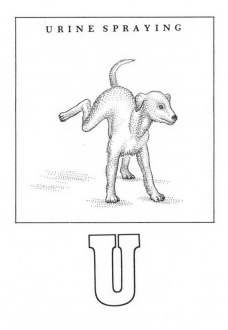

URINE SPRAYING

U

Ultimate cause   Ultimate factor; ultimate function; selective factor. The contrasting terms "ultimate factor (or cause)" and "proximate factor (or cause)" originated in the ornithological literature, where they were introduced by Baker (1938) in connection with the environmental factors that regulate the timing of breeding in birds. Like most animals, birds restrict their breeding, as a rule, to a specific season (see ANNUAL PERIODICITY). Their breeding periods are so timed that the young leave the nest when living conditions are most propitious. The effect of environmental conditions—the paramount one being food supply—is that individuals breeding too early or too late leave fewer offspring than individuals breeding at the optimal time, so the hereditary influences responsible for the poorer timing are gradually eliminated from the population. Such conditions are called ultimate factors because they exercise their effect on the timing of breeding indirectly in the course of generational succession rather than directly via reproductive physiology.

Because of the length of reproductive cycles, especially in larger species, the ultimate factor "optimal food supply" materializes too late in the season to be a signal for the beginning of the cycle, which includes maturation of the gonads, nest building, courtship and mating, incubation, and so forth. Therefore anticipatory signals are necessary, which affect the organism directly and so are called proximate factors. For birds of the higher latitudes, the most important proximate factor for the beginning of the breeding cycle is the increase of day length (photoperiod) in late winter and spring.

After the distinction was introduced, it was soon recognized that such a difference between selective factors acting phylogenetically and regulative factors acting immediately applies to other behavioral phenomena as well. Consequently ultimate causation now embraces all environmental conditions that lead to longer life expectancy and greater reproductive success for those individuals possessing characteristics best suited to those conditions and so to the progressive spreading of such characteristics through the species or population. In this way predator pressure or a need for foraging efficiency can lead to the evolution of group formation. Similarly, food supply, prospects for concealment, and other prerequisites determine which habitats are best for the breeding of a species or population, and those individuals that select them are favored.

On the other hand, "proximate" in this context applies to those factors that actually induce group cohesion and the seeking of a suitable habitat. In the first case these factors are stimuli emanating from group members and the bonding mechanisms that bring about communal living; in the second case they are the distinguishing characteristics of the habitat to which individuals are tuned and a genetically determined or acquired (say by habitat imprinting) preference for those characteristics. A nice discrimination between ultimate and proximate control can help elucidate the constitution of behavior.

Umwelt   See Environment

Unconditioned reflex   See Reflex

Unconditioned stimulus   See Conditioning

Urge   See Drive; Motivation

Urine Spraying Urine marking. In contrast to ordinary urination, this is a forceful ejection of urine that occurs in some mammals. Depending upon the species or the sex, the urine stream can be directed forward or to the rear. In most species—especially in territorial marking—inanimate objects are the targets, such as exposed places in the territory or on its borders. However, especially in rodents and hares, the urine may also be sprayed on other individuals (see SCENT MARKING). It is then a component of social interaction. In the majority of species, only males spray. The functions of urine spraying are not fully understood in all cases and probably differ between species. The following possibilities have been suggested: to induce reproductive receptivity in females, as a component of courtship; to establish and maintain rank order; and to mark territory. It may be used for defense against conspecifics, in which case females also may spray, for example, to repel the sexual advances of males when sexual arousal is low. In many guinea pigs and porcupine rodents, the urine spraying is accompanied by an upright body posture and is consequently visually quite striking.

322

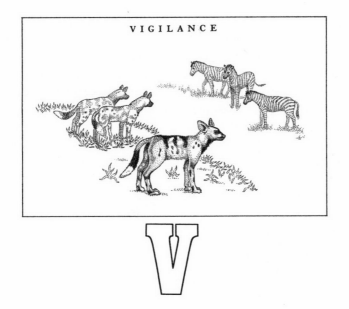

VIGILANCE

V

Vacuum activity Spontaneous performance of a behavior pattern in the absence of the stimuli that normally elicit it. It reflects an extreme lowering of the threshold to stimulation (see THRESHOLD CHANGE). Methodologically, vacuum activities are difficult to identify, because it can almost always be argued that the animal might be reacting to stimulation unnoticed by the observer. But even though incontestably proven cases of "pure" vacuum activities are hard to find, there is no disputing the observations that the susceptibility to release increases with the passing of time since last performance; this was the essential point for ethological theories of instinctive motivation to which the term was most often applied. Probably the most frequently observed case is that of weaver birds performing their complex nest-building movements without nest material or even a substitute object.

Vestigial behavior See Relict

Vigilance Alertness; readiness to detect events that could be of serious concern to an animal or its companions. The term is most often applied to animals exposed to predation. For example, foraging birds of many species periodically break off from their feeding to scan the surroundings for signs of danger. The frequency of such breaks tends to be greatest for birds feeding alone and decreases as group size increases. In some mammal groups particular individuals adopt a sentinel role, looking out for predators while the rest forage uninterruptedly unless the sentinal gives the alarm (see WARNING BEHAVIOR). Parents may show vigilance for the safety of their young, which may not be capable of reliably distinguishing dangerous from harmless situations.

Vigilance can also be observed in the pursuit of food. For instance, predators such as hovering falcons or lurking cats are alert for the appearance of prey, and even a seed-seeking finch is attentive to what corresponds with its search image. Social vigilance means that an animal is alert for signs of possible sexual or aggressive interaction with others, or for location signals to ensure that it does not lose contact with its group.

Physiological monitoring has shown that vigilant animals are usually in a state of heightened arousal, increasing the probability of stimulus detection. The detection of something strange or unexpected can itself increase alertness, as in the case of the orienting response.

Vocalization   Sound produced by the passage of air, causing vibration of membranes in the respiratory tract, such as the vocal cords in the syrinx of birds and the larynx of mammals. Apart from human speech, the most complex vocalizations are the songs of songbirds, whales, and some primates (howler monkeys and gibbons).

Vocal mimicry   A term used in the bioacoustical literature for a species' habit of adding to its vocal repertoire sounds imitated from another species (see IMITATION; MIXED SINGER). Birds kept as pets by people, especially various parrots and mynah birds, can also extend imitation to human speech and inanimate noises. Differing views on the biological significance of vocal mimicry have been expressed. In principle the vocalizations may be addressed to members of the imitated species, to conspecifics, or to other species. In the first case it might be in the service of interspecific territoriality. This appears not to be a general explanation, however, for the imitated species usually shows little or no reaction to the imitations, which often differ in timbre. If the mimicked sounds are directed to conspecifics, they may simply serve as motifs for increasing a bird's song repertoire. Finally, according to the Beau Geste hypothesis, vocal mimicry may give a false indication of higher population density to both conspecifics and members of other species in competition with the singer.

Vocal repertoire   Sound repertoire. Inventory of the sound patterns produced by an individual or a species. Among the most notable for their size and variety of content are the vocal repertoires of many primates (especially chimpanzees and New World monkeys), whales, and birds. In birds that sing, the song repertoire is a subset of the vocal repertoire.

Vocal simulation   Voice simulation; sound simulation. The playback of natural or artificial sounds, as a rule by tape recorder, to test their effects on the behavior of animals being studied for auditory responsiveness. By manipulating the sounds (transposing components in sound sequences, changing rhythm, filtering out certain frequencies), one can determine, from the reactions of the test animals, which characteristics of a naturally occurring sound are significant to its effects and which are not. For example, in species that recognize individuals by their sounds, one can find which characteristics identify a caller individually to those who recognize it. The approach can also be used to assess the hearing capacities of a species—frequency range, amplitude thresholds, discrimination limits, and tuning to specific patterns of frequency or amplitude modulation.

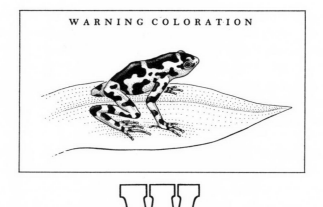

W

Waggle dance   See Dance language

Warning behavior   Alarm behavior. Behavior performed in response to the appearance of a predator that alerts other individuals to the presence of the threat. The most widespread forms are acoustic warning signals, such as the warning and alarm calls of many birds, and the whistling and screaming sounds and noises made by thumping the ground with feet or tail of many mammals (see SOUND PRODUCTION). Warning sounds often have acoustic properties that make it difficult for a predator to locate their source. Chemical warning signals, sometimes referred to as alarm substances, are found in some fish and social insects (see PHEROMONE). There are also visual warning signals, such as the white tail flashed by a fleeing rabbit, and some fish use tactile means to convey alarm. The mobbing that many birds and some mammals direct toward predators can be considered a form of warning behavior.

Warning behavior is often cited as a classic example of altruism, for the giver of the alarm appears to derive no benefit from the action, while putting itself at risk by drawing the attention of the predator. However, data are wanting on precisely how seriously the alarm giver is endangered. Moreover, recent findings and conjectures argue that, in addition to the indirect promotion of inclusive fitness associated with much behavior described as altruistic, warning behavior may also be of more direct advantage to the alarm giver. For example, if it contributes to a joint chorus of warnings, the combined effect may cause the predator to leave the vicinity. There is also the possibility that a warning signal is not directed primarily at conspecifics but at the predator, as if to inform it: "I have seen you, so you might as well stop wasting your time."

Warning coloration   Aposematism. A general term for the typically conspicuous body coloration of ferocious, venomous, or foul-tasting animals, which advertises these characteristics to a prospective predator and thus serves a protective function. Besides the "genuine" form, there is also false warning coloration: many species lacking the repellent attributes have, in the course of their evolution, acquired the appearance or other characteristics of species with such attributes and so are also avoided by predators. This imitation is referred to as Batesian mimicry.

Weaning   The breaking of an offspring's dependence on its mother or parents, especially with regard to feeding. Often the young will persist in trying to get fed after the parent begins to hold back, and a weaning conflict arises (see PARENT-OFFSPRING CONFLICT). In the higher mammals, weaning is sometimes a lengthy process, involving behavioral changes on both sides, with the parent taking the active part as a rule.

In many species of birds, especially those in which the juveniles assist in the care of succeeding broods, and also in many mammals, the young remain with the family beyond the time of weaning, that is, after becoming nutritionally independent. Among other possibilities, this staying together may provide formative social experiences for the young (see TRADITION).

Weaning conflict   See Parent-offspring conflict

Whitten effect   See Pheromone

Wisconsin General Test Apparatus (WGTA) An experimental apparatus developed by Harlow (1949) for the investigation of cognition in primates. The experimental animal is placed in a cage from which it can reach out to uncover one of two or more containers presented to it on a sliding tray. One of the containers will contain a reward item, and the lids of the containers will be differentially marked with colors or symbols with which the reward can be associated. Between the choice trials the animal is prevented from seeing the tray by an opaque partition, while the experimenter arranges the position of rewards and discriminanda. The animal's behavior is observed through a one-way screen. By combining the lid markings in various ways, the experimenter can test the animal's sensory discrimination capacities or its capacity for concept formation.

YAWNING

Yawning   See Comfort behavior

Z

Zeitgeber Time giver. An external stimulus that brings an animal's endogenous periodicity (a diurnal or annual rhythm, say) into synchrony with environmental periodicity. The most important diurnal zeitgeber for most organisms is the light-dark cycle (see CIRCADIAN RHYTHM). For annual periodicity (circannual rhythm) the lengthening of the days in spring is the common zeitgeber governing the onset of reproductive activity.

Within the "proximate factor/ultimate factor" dichotomy the concept of the zeitgeber belongs with the proximate factors; indeed at one time the two were used more or less synonymously. However, with the extension of the proximate/ultimate dichotomy to nontemporal regulating factors, the zeitgeber is now included in a wide range of proximate causes. Moreover, even in the temporal domain some changes are regulated solely by the environment, so the current preference is to restrict use of the term "zeitgeber" to external synchronization factors affecting rhythms whose periodicities are basically endogenous.

Zoo biology A branch of biology, including part of applied ethology, concerned with the biology of zoo animals. Among other things, this study can lead to a better understanding of many behavioral characteristics of animals; applying such knowledge to the layout of free-ranging enclosures and cages or to appropriate feeding and maintenance practices can contribute to appropriately naturalistic conditions for zoo animals.

Behavioral observations of zoo animals have recently taken on added significance, especially for species threatened with extinction in their natural state; by investigating the behavior of these species, scientists can determine the prerequisites for their rehabilitation in the wild or effective maintenance and breeding in captivity.

In spite of the usual disadvantages of working with caged animals, observation of zoo animals offers some specific advantages, including the possibility of detailed long-term study of individual differences in behavior and charting of a species' range of variation in behavior, which may not be fully expressed under feral conditions. Zoo observations have shown that species that live as solitary animals in nature are capable of forming a social rank order when housed in groups, a characteristic otherwise found only in group-living species. Useful knowledge about other kinds of behavior, such as food begging and exploratory behavior, can be gained from the study of zoo animals.

Going beyond the confines of behavioral science, zoo biology is understood to be "the science concerned with all those phenomena that occur in the zoological garden and—in the widest sense—are of biological significance" (Hediger 1941). Zoo biology is therefore a sort of border territory on mixed terrain, including parts of pathology, veterinary medicine, and human medicine, as well as zoological systematics and taxonomy; for example, new geographical races have been "discovered" and described in zoo animals, and their systematic relationships worked out.

Zoomorphism The reverse of anthropomorphism; instead of "humanizing the brutes," zoomorphism "brutalizes humans" (Broadbent 1961). Human behavior is viewed in the light of animal behavior, especially the patterns and strategies of social life, which are thus ascribed more to biological constraints and imperatives than to ethical or rational considerations. Some treatments of human behavior by sociobiology and human ethology have been criticized as zoomorphic.

Zoosemiotics See Communication; Semiotics.

BIBLIOGRAPHY

Aschoff, J. 1954. Zeitgeber der tierischen Tagesperiodik. *Die Naturwissenschaften* 41: 49–56.

Baker, J. R. 1938. The evolution of breeding seasons. In *Evolution: essays on aspects of evolutionary biology*, ed. G. R. de Beer. Oxford: Clarendon Press.

Barlow, G. W. 1968. Ethological units of behavior. In *The central nervous system and fish behavior*, ed. D. Ingle. Chicago: University of Chicago Press.

——— 1977. Modal action patterns. In *How animals communicate*, ed. T. A. Sebeok. Bloomington: University of Indiana Press.

Barnett, S. A. 1981. *Modern ethology: The Science of animal behavior*. New York: Oxford University Press.

Berthold, P. 1971. Physiologie des Vogelzuges. In *Grundriss der Vogelzugskunde*, ed. E. Schüz. Berlin: Paul Parey.

Brentano, F. 1874. *Psychologie vom empirischen Standpunkte*. Leipzig: Dunker and Humbolt.

Broadbent, D. E. 1961. *Behaviour*. London: Eyre and Spottiswoode.

Bruce, H. M. 1961. The pregnancy-block induced in mice by strange males. *Journal of Reproduction and Fertility* 2: 138–142.

——— 1967. Effects of olfactory stimuli on reproduction. In *The effects of external stimuli on reproduction*, ed. G. E. W. Wolstenholme and M. O'Connor. London: Churchill Livingstone.

Chance, M. R. A. 1962. An interpretation of some agonistic postures: The role of "cut-off" acts and postures. *Symposium of the Zoological Society of London* 8: 71–89.

Curio, E. 1969. Funktionsweise und Stammesgeschichte des Flugfienderkennens einiger Darwinfinken (Geospizinae). *Zeitschrift für Tierpsychologie* 26: 394–487.

——— 1976. *The ethology of predation*. Berlin: Springer Verlag.

——— 1978. The adaptive significance of avian mobbing. 1. Teleonomic hypotheses and predictions. *Zeitschrift für Tierpsychologie* 48: 175–183.

Darling, F. F. 1938. *Bird flocks and the breeding cycle*. Cambridge: Cambridge University Press.

333

BIBLIOGRAPHY

Darwin, C. 1859. *On the origin of species by natural selection or the preservation of favoured races in the struggle for life.* London: John Murray.

―――― 1871. *The descent of man and selection in relation to sex.* London: John Murray.

―――― 1872. *The expression of the emotions in man and animals.* London: John Murray.

Dawkins, R. 1976. *The selfish gene.* Oxford: Oxford University Press.

Dawkins, R., and J. R. Krebs. 1976. Animal signals: Information or manipulation? In *Behavioural ecology: An evolutionary approach,* ed. J. R. Krebs and N. B. Davies. Oxford: Blackwell.

Eldridge, N., and S. J. Gould. 1972. Punctuated equilibria: An alternative to phyletic gradualism. In *Models in paleobiology,* ed. T. Schopf. San Francisco: Freeman Cooper.

Eshkol, N., and A. Wachmann. 1958. *Movement notation.* London: Weidenfeld and Nicolson.

Hamilton, W. D. 1964. The genetical evolution of social behaviour. *Journal of Theoretical Biology* 7: 1–52.

―――― 1975. Innate social attitudes of man: An approach from evolutionary genetics. In *Biosocial anthropology,* ed. R. Fox. New York: Wiley.

Harlow, H. F. 1949. The formation of learning sets. *Psychological Review* 56: 51–65.

Hebb, D. O. 1949. *The organization of behavior.* New York: Wiley.

Hediger, H. 1941. Biologische Gesetzmässigkeiten in Verthalten von Wirbeltieren. *Mitteilungen der Naturforschenden Gesellschaft Bern* (1940).

James, W. 1890. *The principles of psychology.* New York: Henry Holt.

Kohler, W. 1925. *The mentality of apes.* New York: Harcourt Brace.

Kortlandt, A. 1940. Wechselwirkung zwischen Instinkten. *Archives neerlandaises de Zoologie* 4: 442–520.

Kummer, H. 1971. *Primate societies.* Chicago: Aldine.

Lorenz, K. 1937. Über den Begriff der Instinkthandlung. *Folia biotheoretica* 2: 17–50.

―――― 1941. Vergleichende Bewegundstudien an Anatinen. *Journal für Ornithologie* 89: 194–293.

―――― 1950. The comparative method in studying innate behaviour patterns. *Symposium of the Society for Experimental Biology* 4: 221–268.

―――― 1981. *The foundations of ethology.* New York: Springer Verlag.

McFarland, D. J. 1971. *Feedback mechanisms in animal behaviour.* New York: Academic Press.

―――― 1974. Time sharing as a behavioral phenomenon. *Advances in the Study of Behavior* 5: 201–225.

―――― 1985. *Animal Behavior.* Menlo Park, Calif.: Benjamin-Cummings.

Markl, H. 1976. *Aggression und Altruismus.* Constance: Konstanz Universiteitsverlag.

Maynard Smith, J. 1974. The theory of games and the evolution of animal conflicts. *Journal of Theoretical Biology* 47: 209–211.

Mayr, E. 1963. *Animal species and evolution.* Cambridge, Mass.: Harvard University Press.

Miller, G. A., E. Galanter, and K. H. Pribram. 1960. *Plans and the structure of behavior.* New York: Henry Holt.

Mock, D. W. 1984. Siblicidal aggression and resource monopolization in birds. *Science* 225: 731–733.

334

Morgan, C. L. 1894. *An introduction to comparative psychology*. London: Scott.

Morris, C. W. 1946. *Signs, language and behavior*. New York: Braziller.

Morris, D. 1957. "Typical intensity" and its relation to the problem of ritualization. *Behaviour* 11: 1–12.

Noble, G. K. 1939. The role of dominance in the social life of birds. *Auk* 56: 263–273.

Orians, G. 1961. Social stimulation within blackbird colonies. *Condor* 63: 330–337.

Osche, G. 1973. *Ökologie. Grundlagen-Erkenntnisse, Entwicklung der Umweltforschung*. Freiburg: Herder.

Parker, G. A. 1974. Courtship persistence and female guarding as male time investment strategies. *Behaviour* 48: 157–184.

Rensch, B. 1954. *Neuere Probleme der Abstammungslehre*. Stuttgart: Enke.

Romanes, G. J. 1882. *Animal intelligence*. London: Kegan Paul, Tench.

Schleidt, W. M. 1964. Über die Spontaneität von Erbkoordinationen. *Zeitschrift für Tierpsychologie* 21: 235–256.

Schneirla, T. C. 1949. Levels in the psychological capacities of animals. In *Philosophy for the future*, ed. R. W. Sellars, V. J. McGill, and A. M. Farber. New York: Humanities Press.

————— 1965. Aspects of stimulation and organization in approach/withdrawal processes underlying vertebrate behavioral development. *Advances in the Study of Behavior* 1: 1–74.

Schöne, H. 1973. Raumorientierung, Begriffe und Mechanismen. *Fortschrift der Zoologie* 21: 1–19.

Sherrington, C. S. 1906. *The integrative action of the nervous system*. Cambridge: Cambridge University Press.

Skinner, B. F. 1938. *The behavior of organisms*. New York: Appleton-Century-Crofts.

————— 1959. A case study of scientific method. In *Psychology: A study of a science*, vol. 2, ed. S. Koch. New York: Appleton-Century-Crofts.

————— 1963. Operant behavior. *American Psychologist* 18: 503–515.

Smith, W. J. 1965. Message, meaning, and context in ethology. *American Naturalist* 99: 405–409.

Teuber, H. L. 1960. Perception. In *Handbook of physiology*. Vol. 3: *Neurophysiology*, ed. J. Field. Washington: American Physiological Society.

Thorndike, E. L. 1911. *Animal intelligence*. 2d Ed. New York: Macmillan.

Tinbergen, N. 1951. *The study of instinct*. Oxford: Oxford University Press.

————— 1959. Comparative studies of the behaviour of gulls (Laridae): a progress report. *Behaviour* 15: 1–70.

Tolman, E. C. 1948. Cognitive maps in rats and men. *Psychological Review* 55: 189–208.

Trivers, R. L. 1971. The evolution of reciprocal altruism. *Quarterly Review of Biology* 46: 35–57.

————— 1972. Parental investment and sexual selection. In *Sexual selection and the descent of man, 1871–1971*, ed. B. Campbell. Chicago: Aldine.

————— 1974. Parent-offspring conflict. *American Zoologist* 14: 249–264.

Uexküll, J. von. 1909. *Umwelt und Innenwelt der Tiere*. Berlin: Springer Verlag.

————— 1957. A stroll through the worlds of animals and men. In *Instinctive behavior: The development of a modern concept*, ed. and trans. C. H. Schiller. New York: International Universities Press.

BIBLIOGRAPHY

Watson, J. B. 1913. Psychology as the behaviorist views it. *Psychological Review* 20: 158–177.

Weidmann, U. 1958. Verhaltensstudien an der Stockente (*Anas platyrhyncos* L.). II. Versuche zur Auslösung und Prägung der Nachgehend Auschluss-reaktion. *Zeitschrift für Tierpsychologie* 15: 277–300.

Wickler, W. 1980. Vocal duetting and the pair bond. I. Coyness and partner commitment, a hypothesis. *Zeitschrift für Tierpsychologie* 52: 201–209.

Wickler, W., and U. Seibt. 1977. *Das Prinzip Eigennutz*. Hamburg: Hoffmann and Campe.

Wilson, E. O. 1975. *Sociobiology: The new synthesis*. Cambridge, Mass.: Harvard University Press.

Wynne-Edwards, V. C. 1962. *Animal dispersion in relation to social behaviour*. Edinburgh: Oliver and Boyd.